LONDON MATHEMATICAL SOCIETY STUDEN_ _____

Managing editor: Dr C.M. Series, Mathematics Institute
University of Warwick, Coventry CV4 7AL, United Kingdom

London Mathematical Society Student Texts 4

An Introduction to Twistor Theory

Second Edition

S. A. Huggett
University of Plymouth

K. P. Tod
University of Oxford

CAMBRIDGE
UNIVERSITY PRESS

Published by the Press Syndicate of the University of Cambridge
The Pitt Building, Trumpington Street, Cambridge CB2 1RP
40 West 20th Street, New York, NY 10011-4211, USA
10 Stamford Road, Oakleigh, Melbourne 3166, Australia

First edition published 1985
Second edition first published 1994

Printed in Great Britain at the University Press, Cambridge

Library of Congress cataloging in publication data available

A catalogue record for this book is available from the British Library

ISBN   0 521 45157 4 hardback
ISBN   0 521 45689 4 paperback

To our parents

Contents

Preface

This book is an introduction to Twistor Theory and modern geometrical approaches to space-time structure at the graduate or advanced undergraduate level. The choice of material presented has evolved from graduate lectures given in London and Oxford and we have aimed to retain the informal tone of those lectures.

Topics covered include spinor algebrea and calculus; compactified Minkowski space; the geometry of null congruences; the geometry of twistor space; an informal account of sheaf cohomology sufficient to describe the twistor solution of the zero rest mass equations; the active twistor constructions which solve the self-dual Yang–Mills and Einstein equations; and Penrose's quasi-local-mass constructions. Exercises are included in the text and after most chapters.

The book will provide graduate students with an introduction to the literature of twistor theory, presupposing some knowledge of special relativity and differential geometry. It would also be of use for a short course on space-time structure independently of twistor theory. The physicist could be introduced gently to some of the mathematics which has proved useful in these areas, and the mathematician could be shown where sheaf cohomology and complex manifold theory can be used in physics.

It is a pleasure to acknowledge many useful discussions, comments and corrections from colleagues in London, Oxford and elsewhere. In particular a debt is due to Mike Eastwood, Andrew Hodges, Lane Hughston, Ted Newman, David Robinson, George Sparling, Richard Ward, Ronny Wells and Nick Woodhouse; we are grateful to Val Willoughby for typing the manuscript and dealing patiently with innumerable revisions; but above all we are indebted for teaching and inspiration to Roger Penrose.

Preface to the second edition

In the eight years since the first edition of this book was published the literature of twistor theory has increased very substantially. There have been several books published whose subject matter overlaps to a greater or lesser extent with ours: Ward and Wells (1990) contains a mathematically more thorough and rigorous account of sheaf cohomology and the machinery from complex analysis necessary for twistor theory than we have found space for; Baston and Eastwood (1989) is a mathematical account of the 'Penrose transform' and its generalisation as part of representation theory; Bailey and Baston (1990) is a collection of commissioned reviews which go on in many directions from topics which we touch on more briefly; finally, Mason and Hughston (1990) follows on from Hughston and Ward (1979) by being a collection of articles which had previously appeared in Twistor Newsletter.

None of these precisely duplicates our book and so we have prepared a new edition taking account of some of the developments of the past eight years. The principal changes for the second edition are as follows:

- chapter 9 has been rewritten with a slightly different emphasis. Our intention is to provide a much clearer and more detailed motivation for the notion of a sheaf, with readers new to the subject in mind. Then (as in the first edition) we go on to discuss the elements of Čech cohomology.

- chapter 13 has been extended to describe the present status of the original quasi-local mass construction. There has been a lot of work in this area, and also a proliferation of other definitions of quasi-local mass. We describe one of them, that of Dougan and Mason, in detail.

- chapter 14 is new. It describes the cohomological framework for studying the kind of multi-linear functionals of zero rest mass fields which arise naturally in twistor theory, and which provide a starting point for a twistor approach to quantum field theory.

- chapter 15, the previous chapter 14, has been extended to describe the further 'Further Developments' of the past eight years. These are impressive (even if some of twistor theory's original aims have not yet been fulfilled). Perhaps the browsing mathematical physicist could do worse than start here!

- chapter 16 is new. Part of our aim in the first edition was to smuggle in extra material disguised as exercises, some of which were perhaps unreasonably difficult. With the inclusion of 'Hints, Solutions and Notes to the exercises' we continue this policy but with what we hope is a more accessible set of exercises.

The final major change is that the second edition has been completely reset in LAT$_E$Xand we are very grateful to Domonic Green for undertaking this substantial labour.

Chapter 1

Introduction

Twistor Theory began as a subject in the late 1960's with the appearance of Penrose's two papers (1967, 1968a). A more definitive statement of its aims and accomplishments was 'Twistor Theory: An Approach to the Quantisation of Fields and Space-time' (Penrose and MacCallum), which appeared in 1973.

What is apparent from the start is the breadth of application which Roger Penrose saw for it. In the decades since then the subject has grown in different direction to the extent that different traditions have emerged.

There is a purely mathematical strain which refers to the Penrose Transform and is interested in its geometrical and complex analytic features, frequently in a positive definitve setting.

There is a quantum field theoretic strain concerned with elementary particles and their interactions in Minkowski space.

There is a modest point of view which simply holds that the theory is useful for solving some non-linear equations and the object is to discover which ones, and there is a full-blooded strain which holds that the repeated occurence and usefulness of complex numbers and complex analyticity tells one something fundamental about the physical world.

The full-blooded strain of twistor philosophy may be seen in for example (Penrose 1975). One place where the diversity of twistor theory is manifest is the Twistor Newsletter, an informal publication produced by the Oxford group about twice a year. The content of the first ten Newsletters was published as 'Advances in Twistor Theory' (Hughston and Ward 1979) and later articles were collected in 'Further Advances in Twistor Theory' (Mason and Hughston 1990).

In this book our aim is simply to give an introduction to the subject and point the reader in the direction of these other possibilities. One thing we do want to emphasize is that a large part of the material falls under the heading of space-time geometry and indeed a course on that subject could be made out of the first six chapters.

The plan of the book is as follows.

We begin in chapter 2 with a review of tensor algebra and calculus as a reminder and to fix conventions. Lorentzian spinors at a point are introduced in chapter 3 and spinor algebra is developed. Also the definition of a complex manifold is given with projective spinors as a paradigm. In chapter 4 we define spinor bundles and the spinor covariant derivative and introduce various spinor differential equations, notably the zero rest mass free field equations and the twistor equation.

In attempting to define a Lie derivative of spinors, we are led in chapter 5 to a consideration of compactified Minkowski space and conformal invariance.

The geometry of null geodesic congruences is discussed in terms of spinors in chapter 6 and the connection between shear and complex analyticity is noticed. We find shear-free congruences in terms of free analytic functions of three variables.

Twistors are introduced in chapter 7, first as spinor fields solving the twistor equation and acted on by the conformal group. The geometrical correspondence between twistor space and complexified compactified Minkowski space is developed and other pictures of a twistor, as an α-plane or a Robinson congruence are given.

In chapter 8 we give the twistor contour integral solution of the zero rest mass free field equations and in attempting to understand curious features of the solution we are led in chapter 9 to sheaf cohomology. After an informal account of sheaf cohomology we return in chapter 10 to the zero rest mass equations and interpret the contour integrals cohomologically.

In chapters 11 and 12 we describe two 'active' constructions where field equations in space-time are coded into deformations of complex structure in corresponding twistor spaces. These are the construction due to Ward for solving the self-dual Yang–Mills equations and the construction due to Penrose for solving the self-dual Einstein equations.

An application of twistor theory in conventional general relativity is discussed in chapter 13. This is Penrose's proposal for a quasi-local momentum-angular-momentum in an arbitrary curved space-time.

The sheaf cohomology of chapter 9 is extended in chapter 14, in which we show how to describe some spaces of multilinear functionals of zero rest mass fields.

Finally in chapter 15 we briefly mention some other developments in twistor theory not mentioned elsewhere!

We have provided exercises throughout, both within the text and grouped at the end of the chapters and have included a chapter (16) of hints, solutions and notes. Some of the exercises are just problems on the material covered but others are open-ended and intend to lead the reader into regions which we don't have the space to cover more fully. (Some of *these* exercises are then referred to later in the text!)

Chapter 2

Review of Tensor Algebra and Calculus

This chapter is chiefly intended as a reminder and to fix conventions (which largely follow Penrose and Rindler 1984).

We shall be concerned with a real four-dimensional vector space V, its dual V^* and its complexification $V_{\mathbf{C}} = V \otimes \mathbf{C}$.

Vectors or elements of V are written V^a; covectors, elements of V^*, are written W_a and the pairing

$$V \times V^* \to \mathbf{R} \text{ is } (V^a, W_a) \to V^a W_a.$$

We follow the abstract index convention of Penrose (1968b) in which the index a on V^a is simply an indication that the object is a vector, rather than one of a set of numbers.

Higher valence tensors are elements of tensor products of V with V^* as, e.g. :

$$P^{a_1 \ldots a_r}{}_{b_1 \ldots b_s} \in \underbrace{V \otimes \ldots \otimes V}_{r} \otimes \underbrace{V^* \otimes \ldots \otimes V^*}_{s}.$$

Algebraic operations on tensors which we shall require are:

i) Contraction: $P^{ab \ldots c}{}_{de \ldots f} \to P^{ab \ldots c}{}_{ae \ldots f}$

ii) Symmetrisation: $P^{(a \ldots b)} = \frac{1}{r!} \sum_{\sigma} P^{\sigma(a) \ldots \sigma(b)}$

where $P^{a \ldots b}$ has r indices and the sum is over all permutations. For example

$$P^{(ab)} = \frac{1}{2}(P^{ab} + P^{ba}).$$

5

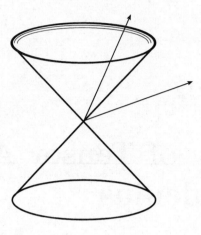

Figure 2.1. The null cone in V separating time-like from space-like.

iii) Skew or anti-symmetrisation: $P^{[a...b]} = \frac{1}{r!} \sum_\sigma (\text{sign}\sigma) P^{\sigma(a)...\sigma(b)}$

where sign σ is ± 1 according as σ is an even or odd permutation. For example

$$P^{[ab]} = \frac{1}{2}(P^{ab} - P^{ba}).$$

A skew-symmetric tensor of valence 2 will be referred to as a *bivector*.

We shall suppose that V comes equipped with a Lorentzian metric, i.e. a symmetric non-degenerate tensor η_{ab} which is equal to diag $(1, -1, -1, -1)$ in an orthonormal frame.

Non-zero vectors in V are characterized as time-like, space-like or null according as their 'length' $\eta_{ab}V^aV^b$ is positive, negative or zero. This gives rise to the characteristic picture of the null cone in V (see figure 2.1).

Removing the origin O disconnects the null cone, so that time-like and null vectors can be further distinguished into two classes, one of which can be labelled future-pointing and the other past-pointing.

The metric allows the identification of V and V^*:

$$V \to V^* \; ; \; V^a \to V_a = \eta_{ab}V^b$$

together with the inverse:

$$V^* \to V \; ; \; V_b \to V^b = \eta^{bc}V_c$$

where $\eta^{ab}\eta_{bc} = \delta^a{}_c$.

Usually a choice of orientation for V is made, that is, a choice of skew tensor $\epsilon_{abcd} = \epsilon_{[abcd]}$ with

$$\epsilon_{abcd}\epsilon^{abcd} = -24.$$

Then a right-handed orthonormal frame (T^a, X^b, Y^c, Z^d) is one with

$$\epsilon_{abcd}T^a X^b Y^c Z^d = 1.$$

With the aid of ϵ_{abcd} and the metric we may define the dual, $*F_{ab}$, of a bivector F_{ab} as:

$$*F_{ab} = -\frac{1}{2}\epsilon_{ab}{}^{cd}F_{cd}.$$

As a consequence of the signature of the metric, we find

$$**F_{ab} = -F_{ab}.$$

Thus the eigenvalues of duality on bivectors are $\pm i$ and therefore the eigen-bivectors are necessarily complex.

The *Lorentz group* $L = O(1,3)$ is the group of endomorphisms of V preserving η_{ab}:

$$\Lambda^a{}_b \in L \Leftrightarrow \Lambda^a{}_b\Lambda^c{}_d\eta_{ac} = \eta_{bd}$$

or, in matrix notation,

$$\eta = \Lambda^T\eta\Lambda \text{ where } \eta = \text{diag}(1, -1, -1, -1).$$

Clearly $\det \Lambda = \pm 1$. Lorentz transformations with negative determinant change the orientation since

$$\epsilon_{abcd}\Lambda^a{}_p\Lambda^b{}_q\Lambda^c{}_r\Lambda^d{}_s = (\det\Lambda)\epsilon_{pqrs}.$$

Also a Lorentz transformation will interchange the future and past null cones if $\Lambda^0{}_0 < 0$. Thus L has four components which may be represented as

$$L = L^\uparrow_+ \cup L^\downarrow_+ \cup L^\uparrow_- \cup L^\downarrow_-$$

where $\pm$ indicates the sign of the determinant and $\uparrow$ indicates $\Lambda^0{}_0 > 0$. The first component $L^\uparrow_+$ contains the identity and is referred to as the *proper orthochronous Lorentz group*.

As examples of Lorentz transformations, we have

i) $\text{diag}(-1,1,1,1) \in L^\downarrow_-$: time-reflection

ii) $\text{diag}(1,-1,1,1) \in L^\uparrow_-$: space-reflection

iii) $\begin{bmatrix} \cosh\psi & 0 & 0 & \sinh\psi \\ 0 & 1 & 0 & 0 \\ 0 & 0 & 1 & 0 \\ \sinh\psi & 0 & 0 & \cosh\psi \end{bmatrix} \in L_+^\uparrow :$ 'boost' in the (03) plane.

The way to calculate the dimension of the Lorentz group is to consider infinitesimal Lorentz transformations: in matrix form

$$\Lambda = I + \epsilon S.$$

Then, to first order in ϵ,

$$\Lambda^T \eta \Lambda = \eta \Rightarrow S^T \eta + \eta S = 0.$$

This gives ten conditions on the sixteen components of S, so that L is six-dimensional.

We assume the reader is familiar with the basic machinery of differential geometry, i.e. the definitions of a smooth manifold and its tangent and cotangent bundles as given for example in Hicks (1965) or Do Carmo (1976, 1992).

We are thinking of the vector space V above as being the tangent space $T_p M = (TM)_p$ at a point p of a real four-dimensional manifold M. The metric on V comes from a metric on M, that is a smooth, valence 2, non-degenerate symmetric tensor field g_{ab} with signature -2. This allows the definition of the orthonormal frame bundle B of M and also determines a unique, torsion-free, metric-preserving connection, the Levi-Civita connection, which we denote ∇_a.

Commuted derivatives give rise to curvature via the Ricci identity:

$$(\nabla_a \nabla_b - \nabla_b \nabla_a) V^d = R_{abc}{}^d V^c \tag{2.1}$$

which defines the Riemann tensor $R_{abc}{}^d$. The Ricci tensor R_{ab} and Ricci scalar R are defined by

$$R_{ab} = R_{acb}{}^c ; \; R = g^{ab} R_{ab},$$

and the Einstein field equations of general relativity are

$$R_{ab} - \frac{1}{2} R g_{ab} = -T_{ab}$$

with a suitable choice of units, where T_{ab} is the stress–energy tensor of matter. The Riemann tensor satisfies the first Bianchi identity:

$$R_{a[bcd]} = 0 \tag{2.2}$$

which implies the interchange symmetry of the Riemann tensor and the symmetry of the Ricci tensor:

$$R_{abcd} = R_{cdab} \; ; \; R_{ab} = R_{ba},$$

and also the *second Bianchi identity* (which is usually referred to as *the* Bianchi identity):

$$\nabla_{[a}R_{bc]de} = 0. \tag{2.3}$$

Finally, the *Lie derivative* along a vector field X^a is defined by

i) $\mathcal{L}_X f = X^a \nabla_a f$ for functions f

ii) $\mathcal{L}_X Y^a = X^b \nabla_b Y^a - Y^b \nabla_b X^a$

iii) for arbitrary valence tensors, extend by the Leibniz rule.

The Lie derivative is actually part of the differential structure of M and is defined prior to the assumption of a metric. Put another way, the above definition is independent of (symmetric) connection.

Our next step is to introduce spinors and develop the algebra and calculus of spinors by analogy with the tensor algebra and calculus.

Exercises 2

a) Given a tensor R_{abcd} with the symmetries of the Riemann tensor:

$$R_{abcd} = R_{[ab]cd} = R_{ab[cd]} \; ; \; R_{a[bcd]} = 0$$

define $S_{abcd} = R_{c(ab)d}$.
Show that S_{abcd} has the symmetries of R_{abcd} but with skew-symmetrisers replaced by symmetrisers (i.e. square brackets replaced by round brackets). Find an expression for R_{abcd} in terms of S_{abcd}.

b) A bivector F_{ab} is said to be simple iff it can be written as the skew outer product of two vectors $2U_{[a}V_{b]}$. Show that F_{ab} simple $\Leftrightarrow \epsilon^{abcd}F_{ab}F_{cd} = 0$
$$\Leftrightarrow {}^*F^{ab}F_{ab} = 0.$$

c) Show that $\mathcal{L}_X g_{ab} = 2\nabla_{(a}X_{b)}$, where ∇_a is the Levi-Civita connection preserving the metric g_{ab}.

d) Show that $\mathcal{L}_X {}^*F_{ab} = {}^*(\mathcal{L}_X F_{ab})$ iff $\mathcal{L}_X g_{ab} = \lambda g_{ab}$ for some function λ. A vector field X^a with this property is known as a conformal Killing vector (see chapter 5).

e) If X^a, Y^a are two conformal Killing vectors show that $\mathcal{L}_X Y^a$ is also.

Chapter 3

Lorentzian Spinors at a Point

Given a vector $V^a \in V$ with components (V^0, V^1, V^2, V^3) in some orthonormal frame, we may define a Hermitian matrix $\Psi(V^a)$ by (Penrose 1974; Penrose and Rindler 1984; Pirani 1965)

$$\Psi(V^a) = V^{AA'} = \begin{bmatrix} V^{00'} & V^{01'} \\ V^{10'} & V^{11'} \end{bmatrix} = \frac{1}{\sqrt{2}} \begin{bmatrix} V^0 + V^3 & V^1 + iV^2 \\ V^1 - iV^2 & V^0 - V^3 \end{bmatrix}. \quad (3.1)$$

The indices A, A' have the ranges $0, 1$ and $0', 1'$ respectively. The significance of the prime will become apparent!

Clearly (3.1) gives a one-one correspondence between 2×2 Hermitian matrices and elements of V. Further, the determinant of the matrix is half the length of the vector:

$$\det \Psi(V^a) = \frac{1}{2} \eta_{ab} V^a V^b.$$

If we multiply the matrix $\Psi(V^a)$ on the left by an element of $SL(2, \mathbf{C})$ (i.e. a 2×2 matrix with complex entries and unit determinant) and on the right by its Hermitian conjugate:

$$V^{AA'} \rightarrow \tilde{V}^{AA'} = t^A{}_B V^{BB'} \bar{t}^{A'}{}_{B'}$$

where $\begin{bmatrix} t^0{}_0 & t^0{}_1 \\ t^1{}_0 & t^1{}_1 \end{bmatrix} \in SL(2, \mathbf{C})$, and

$$\bar{t}^{A'}{}_{B'} = \overline{t^A{}_B},$$

11

then the result will be another Hermitian matrix and the determinant will be unchanged. This process therefore defines a linear transformation on the vector V^a preserving its length, i.e. a Lorentz transformation:

$$V^a \to \tilde{V}^a = \Lambda^a{}_b V^b.$$

Thus we have a map $SL(2, \mathbf{C}) \to L$. The following properties of this map may be readily established:

   i) it is a group homomorphism;

   ii) it is into $L_+^\uparrow$;

   iii) the kernel consists of $\pm I$ in $SL(2, \mathbf{C})$, where I is the identity matrix.

Since $SL(2, \mathbf{C})$ is also a six-parameter group the map is necessarily onto and so is a 2-1 isomorphism. In exercise 3a we establish that $SL(2, \mathbf{C})$ is simply connected, so that this map exhibits $SL(2, \mathbf{C})$ as the universal cover of $L_+^\uparrow$.

The Lorentz transformations in $L_+^\uparrow$ leaving invariant the time-like vector in the chosen orthonormal tetrad evidently define a three-dimensional rotation group, $SO(3)$. The matrix Ψ_0 corresponding to the time-like vector is proportional to the unit matrix, so that the $SO(3)$ subgroup of $L_+^\uparrow$ is covered by an $SU(2)$ subgroup of $SL(2, \mathbf{C})$.

As an example of the map Ψ, the matrix

$$t = \begin{pmatrix} e^{\frac{\psi}{2}} & 0 \\ 0 & e^{-\frac{\psi}{2}} \end{pmatrix}$$

determines a boost in the (03) plane while

$$t = \begin{pmatrix} e^{\frac{i\psi}{2}} & 0 \\ 0 & e^{-\frac{i\psi}{2}} \end{pmatrix}$$

determines a rotation through ψ in the (12) plane.

[A topological aside: a path from I to $-I$ in $SL(2, \mathbf{C})$ will correspond in $L_+^\uparrow$ to a path beginning and ending at I, but which cannot be shrunk to a point. For example

$$t = \begin{pmatrix} e^{\frac{i\lambda}{2}} & 0 \\ 0 & e^{\frac{i\lambda}{2}} \end{pmatrix} \quad 0 \le \lambda \le 2\pi$$

corresponds to a family of rotations in $L_+^\uparrow$ beginning with a zero rotation and ending with a 2π rotation, which is the identity in $L_+^\uparrow$. This path cannot be shrunk to a point in $L_+^\uparrow$. However, if the path in $SL(2,\mathbf{C})$ is continued on to return to I, say by extending the range of λ to $0 \leq \lambda \leq 4\pi$, then *this* path, which is just the previous path traversed twice, can be shrunk to a point in $L_+^\uparrow$. There are a number of ways of demonstrating this fact, all essentially equivalent to the following (see e.g. Misner et al. 1973): imagine a spherical lamp-shade suspended from the ceiling by a long flex and surrounded by notional concentric spheres attached to the flex. Rotate the lamp-shade about a vertical axis and suppose that this rotation is communicated to the concentric spheres so that each is rotated a little less than the one within it, with the outermost one fixed in its original position. Then the various spheres define a set of rotations, i.e. a path in $SO(3)$ from the identity to the rotation undergone by the lamp-shade. If the lamp-shade is rotated through 2π then the flex has a single twist in it and this twist cannot be removed by rearranging the spheres. However, if the lamp-shade is rotated through 4π, then the flex has *two* twists in it and a suitable rearrangement of the spheres will undo this. Essentially one simply takes a loop of flex and passes it below the lamp-shade, but this involves moving all the spheres and shrinking the original path to a point.

Clearly the same argument applies if the lamp-shade is actually connected by any number of strings to different places on the walls, ceiling and floor when the disentangling is more impressive!]

Now we return to the matrices $\Psi(V^a)$ and remark that if the vector V^a is null, then the rank of $\Psi(V^a)$ drops to one. Thus it may be written as the outer product of a complex two-dimensional vector and its complex conjugate:

$$V^{AA'} = \begin{bmatrix} V^{00'} & V^{01'} \\ V^{10'} & V^{11'} \end{bmatrix} = \begin{bmatrix} \alpha^0\overline{\alpha}^{0'} & \alpha^0\overline{\alpha}^{1'} \\ \alpha^1\overline{\alpha}^{0'} & \alpha^1\overline{\alpha}^{1'} \end{bmatrix} = \alpha^A\overline{\alpha}^{A'}. \qquad (3.2)$$

We are therefore led to consider a complex two-dimensional vector space S with elements α^A on which $SL(2,\mathbf{C})$ acts. This is *spin-space* and the elements are *spinors*. The complex conjugate vector space $\overline{S} = S'$ has elements $\beta^{A'}$ and we also have the two dual spaces S^*, S'^* with elements $\gamma_A, \delta_{A'}$.

We may develop the spinor algebra by analogy with the previous section. Higher valence spinors are elements of tensor products:

$$\Phi^{A...BA'...C'}{}_{C...DE'...F'} \in S\otimes...\otimes S\otimes S'\otimes...\otimes S'\otimes S^*\otimes...\otimes S^*\otimes S'^*\otimes...\otimes S'^*$$

We adopt the convention that the relative order of primed and unprimed indices among the upper or among the lower indices is unimportant, i.e.

$$\Phi_{AB'C'} = \Phi_{B'AC'} = \Phi_{B'C'A}.$$

There is an operation of conjugation between S and S': $\alpha^A \in S$ defines $\overline{\alpha}^{A'}$ in S' by $\overline{\alpha}^{A'} = \alpha^A$. This extends to higher valence spinors, e.g. $\alpha^{ABC'} \in S \otimes S \otimes S'$ defines $\overline{\alpha}^{A'B'C} \in S' \otimes S' \otimes S$ by

$$\overline{\alpha}^{A'B'C} = \overline{\alpha^{ABC'}} \text{ so that e.g. } \overline{\alpha}^{0'1'1} = \overline{\alpha^{011'}}.$$

A spinor with equal numbers of primed and unprimed indices, say $\alpha_{ABC'D'}$, is *Hermitian* if

$$\overline{\alpha}_{A'B'CD} = \alpha_{A'B'CD} \text{ so that e.g. } \overline{\alpha}_{0'0'01} = \overline{\alpha_{000'1'}} = \alpha_{0'0'01}.$$

There is a 1-1 correspondence between real tensors of valence n and Hermitian spinors with n primed and n unprimed indices. For $n = 1$, this is the correspondence in (3.2): $V = $ Hermitian part of $S \otimes S'$.

The operations of symmetrisation and anti-symmetrisation are defined for spinors just as for tensors. However, the fact that S is two-dimensional means that any skewing over more than two indices gives zero, and that up to complex multiples there is a unique non-zero skew two-index spinor. We make a choice of one such and call it ϵ_{AB}. Its complex conjugate is written $\epsilon_{A'B'} = \overline{\epsilon}_{A'B'}$, omitting the bar for brevity.

Now for $t_A{}^B \in SL(2, \mathbf{C})$,

$$t_A{}^B t_C{}^D \epsilon_{BD} = (\det t)\epsilon_{AC} = \epsilon_{AC}.$$

Thus the chosen ϵ_{AB} is preserved by $SL(2, \mathbf{C})$ much as the metric η_{ab} is preserved by the Lorentz group L. All spinors are 'null' with respect to ϵ_{AB} in that $\epsilon_{AB}\alpha^A\alpha^B = 0$ for all α^A, and conversely if $\epsilon_{AB}\alpha^A\beta^B = 0$ for non-zero α^A and β^B then they are proportional.

ϵ_{AB} allows the identification of S with S^* via

$$\alpha^A \rightarrow \alpha_A = \alpha^B \epsilon_{BA}$$

and conversely

$$\alpha_A \rightarrow \alpha^A = \epsilon^{AB}\alpha_B$$

where $\epsilon^{AB}\epsilon_{CB} = \delta_C{}^A$. Care is needed with these expressions because ϵ_{AB} is skew. Thus, for example,

$$\alpha_A = \alpha^B \epsilon_{BA} = -\alpha^B \epsilon_{AB}.$$

The mnemonic is 'adjacent indices – descending to the right'. In the same fashion, $\epsilon_{A'B'}$ and $\epsilon^{A'B'}$ identify S' with S'^*. If we choose ϵ_{AB} so that $\epsilon_{01} = 1$ in the basis of S implicitly given by (3.1) and (3.2) then the Hermitian spinor $\epsilon_{AB}\epsilon_{A'B'}$, which corresponds to *some* real tensor, in fact corresponds to the metric η_{ab}. To see this, we first observe that it defines a symmetric, non-degenerate tensor. Next, for any null vector $V^a = \alpha^A \bar{\alpha}^{A'}$ we have

$$\epsilon_{AB}\epsilon_{A'B'}\alpha^A\bar{\alpha}^{A'}\alpha^B\bar{\alpha}^{B'} = (\epsilon_{AB}\alpha^A\alpha^B)(\epsilon_{A'B'}\bar{\alpha}^{A'}\bar{\alpha}^{B'}) = 0$$

so that it is certainly correct up to proportionality. Finally, for the scale, if $t^a = t^{AA'}$ is the unit time-like vector with $\Psi(t^a) = \frac{1}{\sqrt{2}}\begin{bmatrix} 1 & 0 \\ 0 & 1 \end{bmatrix}$ then

$$\epsilon_{AB}\epsilon_{A'B'}t^{AA'}t^{BB'} = \epsilon_{01}\epsilon_{0'1'}t^{00'}t^{11'} + \epsilon_{10}\epsilon_{1'0'}t^{11'}t^{00'} = 1.$$

Since S is two-dimensional, the analogue of a tetrad is a *dyad* (o_A, ι_A) and we can say that this is normalised if $\epsilon_{AB}o^A\iota^B = o_B\iota^B = 1$. A normalised dyad defines a tetrad in V by

$$l^a = o^A\bar{o}^{A'}; n^a = \iota^A\bar{\iota}^{A'}; m^a = o^A\bar{\iota}^{A'}; \overline{m}^a = \iota^A\bar{o}^{A'}. \qquad (3.3)$$

Here l^a and n^a are real future-pointing null vectors and m^a is a complex null vector. With the identification of $\epsilon_{AB}\epsilon_{A'B'}$ with η_{ab} we may also calculate

$$\eta_{ab}l^a n^b = 1 = -\eta_{ab}m^a\overline{m}^b$$

and all other scalar products are zero. Further, the tetrad is right-handed. A tetrad with these properties is referred to as a *null tetrad* and evidently determines a normalised dyad up to a sign ambiguity, i.e. (o^A, ι^A) is not distinguished from $(-o^A, -\iota^A)$. Note that, for a normalised dyad,

$$o_A\iota_B - \iota_A o_B = \epsilon_{AB}. \qquad (3.4)$$

This a particular case of a more general formula. We observe that $\epsilon_{A[B}\epsilon_{CD]}$, being skew on three indices, necessarily vanishes. Writing this out in full, rearranging terms, and raising some indices we find

$$\epsilon_{AB}\epsilon^{CD} = \delta_A{}^C\delta_B{}^D - \delta_A{}^D\delta_B{}^C. \qquad (3.5)$$

Now suppose that a given spinor is skew on a pair of indices, say

$$\Phi_{...CD...} = \Phi_{...[CD]...}$$

where the dots denote some other collection of indices. Then multiplying by the identity (3.5) we find

$$\epsilon_{AB}\epsilon^{CD}\Phi_{...CD...} = 2\Phi_{...AB...}$$

i.e.

$$\Phi_{...AB...} = \frac{1}{2}\epsilon_{AB}\Phi_{...C}{}^C{}_{...}.$$

Thus a skew pair of indices can be removed as an ϵ_{AB} with a contraction. In this sense, *only symmetric spinors count*, that is, any spinor can be written in terms of symmetric spinors and the special spinors $\epsilon_{AB}, \epsilon_{A'B'}$ and their inverses.

A further simplification is that *symmetric spinors factorise* in the sense that a symmetric valence n spinor $\Phi_{A...B}$ can be written as a symmetrised outer product $\alpha_{(A}...\beta_{B)}$ of n valence one spinors. To see this, choose a dyad and consider components; define $\xi^A = (1, x)$ and form $\Phi_{A...B}\xi^A...\xi^B$. This is a polynomial of degree n in x and so, since S is complex, it factorises. The linear factors are of the form $\alpha_A\xi^A, \beta_A\xi^A$ etc. whence $\Phi_{A...B} = \alpha_{(A}...\beta_{B)}$. The individual factors are unordered and defined only up to scale. Each factor defines a null vector via (3.2) and these are known as the *principal null directions* (p.n.d.s) of $\Phi_{A...B}$.

As an example of these properties of spinors, we consider the spinor equivalent of a valence 2 tensor,

$$T_{ab} = T_{AA'BB'} = T_{ABA'B'}.$$

On the unprimed index pair we have

$$
\begin{aligned}
T_{ABA'B'} &= T_{(AB)A'B'} + T_{[AB]A'B'} \\
&= T_{(AB)A'B'} + \frac{1}{2}\epsilon_{AB}T_C{}^C{}_{A'B'}
\end{aligned}
$$

and now on the primed pair

$$
\begin{aligned}
&= T_{(AB)(A'B')} + \frac{1}{2}\epsilon_{A'B'}T_{(AB)C'}{}^{C'} + \frac{1}{2}\epsilon_{AB}T_C{}^C{}_{(A'B')} \\
&\quad + \frac{1}{4}\epsilon_{AB}\epsilon_{A'B'}T_{CC'}{}^{CC'}.
\end{aligned}
\tag{3.6}
$$

Here the first term is a Hermitian spinor and corresponds to the symmetric trace-free part of T_{ab}, the fourth term is a real multiple of the metric and so must be the trace $\frac{1}{4}g_{ab}T_c{}^c$. The middle two terms give a Hermitian spinor when taken together. Further if A and B and also A' and B' are interchanged, each of the two middle terms changes sign. Thus these two give the skew part $T_{[ab]}$ of T_{ab}. We deduce that if F_{ab} is a real bivector then

$$F_{ab} = \Phi_{AB}\epsilon_{A'B'} + \overline{\Phi}_{A'B'}\epsilon_{AB} \tag{3.7}$$

where Φ_{AB} is symmetric.

The spinor Φ_{AB} factorises as $\alpha_{(A}\beta_{B)}$ and the principal null directions are related to eigenvectors of F_{ab} (exercise 3b).

The spinor equivalent of ϵ_{abcd} is found by a similar analysis to be

$$\epsilon_{abcd} = i(\epsilon_{AC}\epsilon_{BD}\epsilon_{A'D'}\epsilon_{B'C'} - \epsilon_{AD}\epsilon_{BC}\epsilon_{A'C'}\epsilon_{B'D'})$$

so that

$$\epsilon_{ab}{}^{cd} = i(\delta_A{}^C\delta_B{}^D\delta_{A'}{}^{D'}\delta_{B'}{}^{C'} - \delta_A{}^D\delta_B{}^C\delta_{A'}{}^{C'}\delta_{B'}{}^{D'}).$$

In this form it is easy to see the effect of dualising the bivector F_{ab} (3.7):

$$F_{ab} = \Phi_{AB}\epsilon_{A'B'} + \overline{\Phi}_{A'B'}\epsilon_{AB}$$
$$^*F_{ab} = \frac{1}{2}\epsilon_{ab}{}^{cd}F_{cd} = -i\Phi_{AB}\epsilon_{A'B'} + i\overline{\Phi}_{A'B'}\epsilon_{AB}.$$

The eigenspaces of dualising are now apparent:

$$\text{if } W_{ab} = \Phi_{AB}\epsilon_{A'B'} \text{ then } {}^*W_{ab} = -iW_{ab} \qquad (3.8)$$
$$\text{if } W_{ab} = \Psi_{A'B'}\epsilon_{AB} \text{ then } {}^*W_{ab} = iW_{ab}. \qquad (3.9)$$

The first case is referred to as an *anti-self-dual* bivector and the second as a *self-dual* bivector. As anticipated, the eigen-objects are complex.

We may exploit this discussion of bivectors to give a geometrical interpretation of a spinor. Firstly, a spinor o^A defines a null vector $l^a = o^A\overline{o}^{A'}$. However, this ignores the phase of o^A, i.e. o^A and $e^{i\theta}o^A$ define the same null vector. To include the information of the phase we define the bivector

$$F_{ab} = o_A o_B \epsilon_{A'B'} + \overline{o}_{A'}\overline{o}_{B'}\epsilon_{AB}. \qquad (3.10)$$

By exercise 2b, this bivector is simple. If we introduce a spinor ι_A to form a normalised dyad with o_A then

$$F_{ab} = o_A o_B(\overline{o}_{A'}\overline{\iota}_{B'} - \overline{\iota}_{A'}\overline{o}_{B'}) + \overline{o}_{A'}\overline{o}_{B'}(o_A\iota_B - \iota_A o_B)$$
$$= l_a m_b - l_b m_a + l_a\overline{m}_b - l_b\overline{m}_a$$

in terms of the null tetrad (3.3) defined by the dyad, so

$$F_{ab} = 2l_{[a}X_{b]}$$

where $X_a = m_a + \overline{m}_a$.

Now X_a is space-like and orthogonal to l_a. A different choice of ι_A, say $\hat{\iota}_A = \iota_A + \lambda o_A$, adds a multiple of l_a to X_a. Thus o_A, as well as defining the null vector l_a, defines a 'null flag' or two-plane element containing the 'flag-pole' l^a (see figure 3.1).

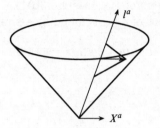

Figure 3.1. The null flag.

If we set $m^a = X^a + iY^a$, then a change of phase $o^a \to e^{i\theta}o^A$ takes $\iota^A \to e^{-i\theta}\iota^A$ and so $m^a \to e^{2i\theta}m^a$. This rotates m^a in the plane of X^a and Y^a by 2θ. Thus as the spinor o^A rotates by θ, the flag-pole rotates by 2θ. As anticipated, there is no way of distinguishing geometrically between $\pm o^A$.

We end this chapter with a consideration of 'spinors up to scale,' i.e. the set of one-dimensional subspaces of the two-dimensional space S'. This is the one-dimensional *complex projective space* $\mathbf{P}^1$ (or $\mathbf{CP}^1$ if we wish to distinguish it from $\mathbf{RP}^1$). We may describe it as follows: a point of $\mathbf{P}^1$ is a proportionality class of primed spinors, an equivalence class $[\pi_{0'}, \pi_{1'}]$ of pairs of complex numbers $(\pi_{0'}, \pi_{1'})$ under the equivalence relation

$$(\pi_{0'}, \pi_{1'}) \sim (\lambda\pi_{0'}, \lambda\pi_{1'}) \; ; \; \lambda \in \mathbf{C}^*,$$

with $\pi_{0'}$ and $\pi_{1'}$ not both zero. We may therefore distinguish two open neighbourhoods which cover $\mathbf{P}^1$:

$$U_0 = \{[\pi_{0'}, \pi_{1'}] : \pi_{0'} \neq 0\}; U_1 = \{[\pi_{0'}, \pi_{1'}] : \pi_{1'} \neq 0\}.$$

On U_0, the quantity $\zeta = \frac{\pi_{1'}}{\pi_{0'}}$ is a good coordinate, i.e. it maps U_0 one-to-one and onto the complex plane. On $U_1, \eta = \frac{\pi_{0'}}{\pi_{1'}}$ is a good coordinate and in the overlap $U_0 \cap U_1, \eta = \zeta^{-1}$.

This exhibits $\mathbf{P}^1$ as a *complex manifold* (see e.g. Morrow and Kodaira 1971; Field 1982). The definition is: a paracompact Hausdorff topological space X is a complex manifold if

a) X has an open cover $\{U_i\}_{i \in I}$ with coordinate functions $f_i : U_i \to \mathbf{C}^n$;

b) on each non-empty intersection $U_i \cap U_j$, the 'transition' functions $f_j \circ f_i^{-1}$ are holomorphic from $\mathbf{C}^n$ to $\mathbf{C}^n$.

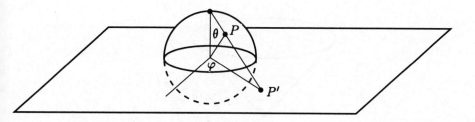

Figure 3.2. Stereographic projection.

The coordinates ζ and η can be given a more geometrical interpretation by means of stereographic projection. $\mathbf{P}^1$ is the space of spinors regardless of scale and phase, and this is the space of null directions or the projective null cone at a point in Minkowski space. Clearly this is topologically a sphere, the celestial sphere. If we represent it as the unit sphere in $\mathbf{R}^3$, then stereographic projection is the map from the north pole to the equatorial plane as in figure 3.2.

This map takes the point p with polar coordinates (θ, φ) to the point p' with

$$\zeta := x + iy = \cot \frac{\theta}{2} e^{i\varphi}$$

and maps the whole sphere with the exception of the north pole onto the Argand plane of ζ. To get the north pole we project from the south pole and the point with polar coordinates (θ, φ) goes to

$$\eta := x - iy = \tan \frac{\theta}{2} e^{-i\varphi}.$$

Here the minus sign is because the equatorial plane when seen from below has the opposite orientation.

Thus U_0 is the sphere minus the north pole, U_1 is the sphere minus the south pole and in the overlap, $\eta = \zeta^{-1}$.

The isomorphism of $SL(2, \mathbf{C})$ with $L_+^\uparrow$ can be seen directly with the aid of these considerations.

The Minkowski metric in spherical polar coordinates is

$$ds^2 = dt^2 - dr^2 - r^2(d\theta^2 + \sin^2 \theta d\varphi^2)$$

and the future null cone N^+ of the origin is just the surface $t = r$. The

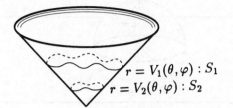

$r = V_1(\theta, \varphi) : S_1$
$r = V_2(\theta, \varphi) : S_2$

Figure 3.3. Two cuts of the future null cone.

degenerate metric on N^+ is therefore

$$ds^2 = -r^2(d\theta^2 + \sin^2\theta d\varphi^2) \tag{3.11}$$

where θ and φ label the null geodesics generating N^+ and r is an affine parameter along them. A particular section or *cut* of N^+ is given by specifying r as a function of θ and φ (see figure 3.3).

It is clear from (3.11) that the map down the generators from S_1 to S_2 is conformal. The cut $\Sigma : r = 1$ is invariantly defined as the set of null vectors l^a with $t^a l_a = 1$ where t^a is the unit vector defining the time axis and the intrinsic metric of Σ is that of a unit sphere. If we apply a Lorentz transformation to t^a, say

$$t^a \to \hat{t}^a = \Lambda^a{}_b t^b$$

then

$$\Sigma \to \hat{\Sigma} : \hat{t}^a l_a = 1$$

and clearly $\hat{\Sigma}$ must also have the unit sphere metric (see figure 3.4).

The map down the generators from $\hat{\Sigma}$ to Σ is conformal so a Lorentz transformation defines an element of the group $C(2)$ of conformal transformations of the unit sphere. If $\Lambda^a{}_b$ is a rotation then $\hat{\Sigma}$ coincides with Σ and the corresponding element of $C(2)$ is a rotation.

Evidently, this is a group homomorphism and is one-one. To go further we need to know more about $C(2)$.

In terms of the coordinate ζ, the unit sphere metric is

$$ds^2 = -\frac{4}{(1 + \zeta\bar{\zeta})^2} d\zeta d\bar{\zeta}. \tag{3.12}$$

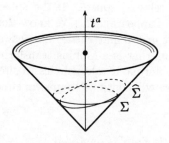

Figure 3.4.

Now we know from the theory of one complex variable (and it is clear from (3.12) – see also exercise 3h) that a conformal map in two dimensions must be holomorphic (or anti-holomorphic, in which case it is orientation-reversing). Thus a conformal transformation of the sphere must be a globally defined holomorphic transformation:

$$\zeta \rightarrow \hat{\zeta} = f(\zeta)$$
$$\eta \rightarrow \hat{\eta} = g(\eta)$$

with $\hat{\eta} = \hat{\zeta}^{-1}$ where finite.

Intuitively, we may find these as follows: the function $f(\zeta)$ must have a simple zero at the transformed south pole and a simple pole at the transformed north pole, so

$$f(\zeta) = \frac{a\zeta + b}{c\zeta + d} h(\zeta)$$

where $h(\zeta)$ has no zeroes or poles. Therefore $h(\zeta)$ must be constant by Liouville's theorem and

$$f(\zeta) = \frac{a\zeta + b}{c\zeta + d}. \tag{3.13}$$

Without loss of generality we may take $ad - bc = 1$. Thus the conformal group $C(2)$ is the group of Möbius transformations (3.13) which is clearly isomorphic to $SL(2, \mathbf{C}) / \{\pm I\}$ and we must have

$$SL(2, \mathbf{C}) \overset{2 \text{–} 1}{\rightarrow} C(2) \cong L_+^\uparrow.$$

This relationship was pointed out in Penrose (1955,1956) 'The Visual Appearance of a Moving Sphere'. A sphere S intersects a circular cone of null geodesics on an observer's past null cone, i.e. S is represented as a circular disc on the celestial sphere. If the sphere moves rapidly past the observer, what is its appearance? (We know that it suffers a 'Lorentz contraction' but does it *appear* flattened?) To answer this question, apply a Lorentz transfomation, that is a Möbius transformation, to the celestial sphere. Since this is conformal it takes circles to circles and hence discs to discs, so that the sphere will continue to *appear* circular!

Exercises 3

a) Show that $SL(2, \mathbf{C})$ is simply connected.

b) Show that if the real bivector F_{ab} corresponds to the symmetric spinor $\Phi_{AB} = \alpha_{(A}\beta_{B)}$, then $\alpha_A\bar{\alpha}_{A'}$ and $\beta_A\bar{\beta}_{A'}$ are eigenvectors of F_{ab}.

c) A *null* bivector F_{ab} is one for which the principal null directions coincide, i.e. $\Phi_{AB} = \alpha_A\alpha_B$. Show that a bivector F_{ab} is null iff

$$F_{ab}F^{ab} = F_{ab}{}^*F^{ab} = 0.$$

Deduce that a null bivector is simple (exercise 2b).

d) If R_{abcd} has the symmetries of the Riemann tensor, show that it has a spinor decomposition of the form

$$
\begin{aligned}
R_{abcd} \quad = \quad & \Psi_{ABCD}\epsilon_{A'B'}\epsilon_{C'D'} + \bar{\Psi}_{A'B'C'D'}\epsilon_{AB}\epsilon_{CD} \\
& + \Phi_{ABC'D'}\epsilon_{A'B'}\epsilon_{CD} + \Phi_{A'B'CD}\epsilon_{AB}\epsilon_{C'D'} \\
& + 2\Lambda(\epsilon_{AC}\epsilon_{BD}\epsilon_{A'B'}\epsilon_{C'D'} + \epsilon_{AB}\epsilon_{CD}\epsilon_{A'D'}\epsilon_{B'C'})
\end{aligned}
$$

where Ψ_{ABCD} is symmetric, $\Phi_{ABA'B'}$ is symmetric on each pair of indices and is Hermitian, and Λ is real. By taking contractions of this, find the spinor equivalents of the Weyl tensor, the Ricci tensor and the Ricci scalar.

e) $\mathbf{CP}^n$ (respectively $\mathbf{RP}^n$) is defined as the set of one-dimensional subspaces of an n+1-dimensional complex (respectively real) vector space. Show that $\mathbf{CP}^n$ is a complex manifold by constructing an appropriate cover and coordinates.

In a similar way, one defines a *Grassmann manifold* $G(k, n)$ as the set of k-dimensional subspaces of an n-dimensional vector space.

f) Recalling that

$$t = \begin{pmatrix} e^{\frac{\psi}{2}} & 0 \\ 0 & e^{-\frac{\psi}{2}} \end{pmatrix} \in SL(2, \mathbf{C})$$

defines a boost in the (03) plane, show that the effect of a boost on the celestial sphere is to move points towards the direction of motion.

g) This exercise is concerned with *(complex) line bundles* over $\mathbf{CP}^1$. Such a bundle, say L, may be defined in terms of the cover $\{U_0, U_1\}$ as follows: there exist local trivialisations

$$t_i : \rho_i L \to U_i \times \mathbf{C}$$

(where $\rho_i L$ means L restricted to U_i), and a transition relation between the fibre coordinates μ_i

$$\mu_0 = f_{01}(\zeta)\mu_1$$

where the transition function f_{01} is holomorphic on $U_0 \cap U_1$.

A *section* of L is defined by a pair of holomorphic functions

$$\mu_0 = s_0(\zeta), \ \mu_1 = s_1(\zeta)$$

related by the transition function on $U_0 \cap U_1$.

We may define a line bundle L over $\mathbf{CP}^1$ by associating to a point $[\pi_{A'}]$ in $\mathbf{CP}^1$ the corresponding one-dimensional subspace of $\mathbf{C}^2$. Then $L \subset \mathbf{CP}^1 \times \mathbf{C}^2$ and the projection $L \to \mathbf{CP}^1$ is

$$\{[\pi_{A'}], \lambda\pi_{A'} : \lambda \in \mathbf{C}\} \to [\pi_{A'}].$$

Show that in this case we may take $f_{01} = \zeta$. This particular L is known as H^{-1} or $\mathcal{O}(-1)$. In a similar way we may define H^{-n} or $\mathcal{O}(-n)$ by

$$\{\lambda \underbrace{\pi_{A'} \ldots \pi_{B'}}_{n} : \lambda \in \mathbf{C}\} \to [\pi_{A'}].$$

Show that now $f_{01} = \zeta^n$. By analogy, we define H^n by $f_{01} = \zeta^{-n}$. Show that sections of H^k are defined by functions $F(\pi_{A'})$ homogeneous of degree k in the sense that

$$F(\lambda\pi_{A'}) = \lambda^k F(\pi_{A'}).$$

Evidently this construction produces line bundles H^n or $\mathcal{O}(n)$ on any complex projective space (Wells 1980).

h) An alternative approach to defining a complex manifold is in terms of extra structure on a real smooth manifold (Chern 1967).

If M is a smooth manifold with tangent bundle TM, form $TM_\mathbf{C}$, the complexification $((TM_\mathbf{C})_p = (TM) \otimes \mathbf{C})$. What is required is a way of recognising vectors in $(TM_\mathbf{C})_p$ which are to be regarded as holomorphic.

A *complex structure* J on a real vector space V is a tensor $J_a{}^b$ with $J_a{}^b J_b{}^c = -\delta_a{}^c$. In $V_{\mathbf{C}}$, J has eigenspaces with eigenvalues $\pm i$. The holomorphic or type $(1,0)$ vectors are defined to be the $+i$ eigenspace and the anti-holomorphic vectors are the $-i$ eigenspace.

An *almost complex structure* on M is a field of complex structures on TM. The almost complex structure is said to be *integrable* if the different definitions of holomorphic at each point fit together. This means that if V^a and U^a are type $(1,0)$ vector fields then their commutator must be also, i.e.

$$(J_a{}^b - i\delta_a{}^b)V^a = 0 = (J_a{}^b - i\delta_a{}^b)U^a \Rightarrow (J_a{}^b - i\delta_a{}^b)(V^c \nabla_c U^a - U^c \nabla_c V^a) = 0.$$

Show that this is equivalent to

$$U^a V^b \nabla_{[a} J_{b]}{}^c = 0$$

or

$$\delta_{[a}{}^c J_{b]}{}^d \nabla_c J_d{}^e = 0$$

and that, for a two-dimensional manifold M, this condition is vacuous.

The Newlander–Nirenberg Theorem then asserts that complex coordinates can be found for M so that M becomes a complex manifold in the earlier sense.

Show that, for an orientable two-dimensional manifold M, having a conformal structure is equivalent to having an almost complex structure. (The point is that type $(1,0)$ vectors will be null.)

Chapter 4

Spinor Fields

By analogy with vector fields, we wish to define a spinor field on a manifold M as a section of a suitable bundle S over M. In chapter 3, we saw that a spinor dyad at a point determines a null tetrad which is a right-handed tetrad containing future-pointing null vectors. To define S therefore the manifold M must be orientable (so that a global choice of orientation is possible) and time-orientable (so that a global choice of future-pointing is possible). This means that the orthonormal frame bundle B of M can be reduced to an $L_+^\uparrow$ bundle, i.e. that we can choose the transition matrices for B to be in $L_+^\uparrow$.

Now a dyad determines a null tetrad but, conversely, the tetrad determines a dyad only up to sign. Thus, having reduced B to an $L_+^\uparrow$ bundle, we must find S as a double cover of B. For this to be possible, M must satisfy certain topological restrictions which can be illustrated diagrammatically (see below) but which we shall first discuss in the language of Čech cohomology (Bott and Tu 1982). Here the sheaves are constant ($\mathbf{Z}_2$) and the reader wishing to follow this argument but unfamiliar with the terminology may want to refer first to chapter 9.

We begin with the question of orientability. We take a locally finite open cover $\{U_i\}_{i \in I}$ of M, and a choice of orthonormal frame f_i over U_i. On the non-empty intersections $U_i \cap U_j$, the frames f_i and f_j are related by a Lorentz transformation P_{ij}:

$$f_j P_{ij} = f_i.$$

(Throughout this discussion, we suspend the summation convention!) The transformations P_{ij} define the orthonormal frame bundle B and must satisfy

$$P_{ji} = P_{ij}^{-1}$$
$$P_{ij}P_{ki}P_{jk} = I \text{ if } U_i \cap U_j \cap U_k \neq \emptyset.$$

We define $\tau_{ij} = \det P_{ij}$.

Then τ_{ij} is an assignment of ± 1 to every non-empty intersection $U_i \cap U_j$ and $\tau_{ij} = \tau_{ji}$. Thus τ_{ij} defines a one cochain $\tau \in C^1(M; \mathbf{Z}_2)$. Further $\tau_{ij}\tau_{ki}\tau_{jk} = 1$ so that τ is actually a cocycle, $\tau \in Z^1(M; \mathbf{Z}_2)$.

A change of orientation of some of the f_i corresponds to a zero cochain $\omega \in C^0(M; \mathbf{Z}_2)$ as follows: $\omega_i = 1$ if the orientation of f_i is unchanged, $\omega_i = -1$ if the orientation of f_i is changed. This in turn modifies the P_{ij} and hence the τ_{ij} according to

$$\tau_{ij} \rightarrow \omega_i \tau_{ij} \omega_j$$

i.e. τ changes by a coboundary.

If the class $[\tau]$ of τ in $H^1(M; \mathbf{Z}_2)$ is trivial then τ is a coboundary, $\tau_{ij} = \omega_i \omega_j$. Now using the cochain ω we can modify the frames f_i so that all the τ_{ij} become one. This is then a choice of orientation – thus the manifold M is orientable iff the class $[\tau]$ in $H^1(M; \mathbf{Z}_2)$ is trivial.

A similar argument shows that M is *time-orientable* (i.e. that a consistent choice of one part of the null cone of each point as the future null cone is possible) if another class is trivial.

Suppose this is done, so that M is orientable and time-orientable and the matrices P_{ij} are all in $L_+^\uparrow$. Now to construct the spin bundle S, we choose an $SL(2, \mathbf{C})$ matrix σ_{ij} which is one of the two inverse images of P_{ij} (the other being $-\sigma_{ij}$). These can evidently be chosen to satisfy

$$\sigma_{ji} = \sigma_{ij}^{-1}$$

but on the non-empty triple intersections we shall have

$$\sigma_{ij}\sigma_{ki}\sigma_{jk} = z_{ijk}I$$

where I is the 2×2 unit matrix and $z_{ijk} = \pm 1$. To be able to construct S we need to be able to choose the σ_{ij} so that all the z_{ijk} are $+1$. Again, z_{ijk} defines a cochain z, now in $C^2(M; \mathbf{Z}_2)$, which again is actually a cocycle. We may define a one-cochain ω_{ij} by changing the choice of $\sigma_{ij} : \omega_{ij} = -1$ if we take the choice $-\sigma_{ij}$, otherwise $\omega_{ij} = 1$. This changes z_{ijk} by a coboundary:

$$z_{ijk} \rightarrow z_{ijk}\tau_{ij}\tau_{ki}^{-1}\tau_{jk}.$$

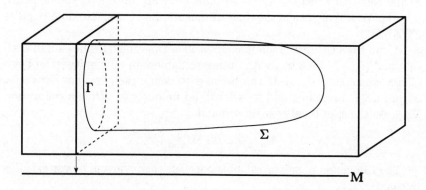

Figure 4.1. Trouble arises if the path Γ can be shrunk to zero over a surface Σ not wholly in the fibre containing Γ.

The class $[z]$ of z_{ijk} in $H^2(M; \mathbf{Z}_2)$ is trivial if z equals a coboundary, say

$$z_{ijk} = \tau_{ij}\tau_{ki}^{-1}\tau_{jk}.$$

In this case we may use the cocycle τ_{ij} to change the signs of the σ_{ij}:

$$\sigma \rightarrow \tau_{ij}\sigma_{ij}.$$

Now the new σ_{ij} satisfy

$$\sigma_{ij} = \sigma_{ji}^{-1} \text{ and } \sigma_{ij}\sigma_{ki}\sigma_{jk} = I$$

and can be used to build the bundle S. The class $[z]$ in $H^2(M; \mathbf{Z}_2)$ is thus the topological obstruction to building S. If $[z]$ *is* trivial there will be in-equivalent bundles S parametrised by one-cocycles with trivial coboundary, i.e. by elements of $H^1(M; \mathbf{Z}_2)$.

These topological considerations may be shown in diagrammatic form as in figure 4.1: the difficulty is that a path in the fibre of the frame bundle not homotopic to zero in that fibre may be homotopic to zero if deformed in the whole frame bundle.

For Γ we imagine a rotation of a frame, at a point p, through 2π. From the discussion in chapter 3, we know that Γ is not homotopic to zero without moving away from p. However, it is possible, and an example is provided by exercise 4a, that if the frame is carried along paths in M beginning and ending at p then this rotation may be undone.

We shall suppose that these pathologies are absent and that the bundle S has been constructed. Local sections of S are unprimed spinor fields, $\pi^A(x)$. We will also have the dual, conjugate, and conjugate dual bundles S^*, S', S'^*.

Again from the discussion in chapter 3, a spinor field defines a null anti-self-dual (a.s.d.) bivector, with an unavoidable sign ambiguity. The Levi-Civita connection ∇_a of M can be used to define the covariant derivative of null a.s.d. bivectors, and so will extend uniquely to define a connection $\nabla_{AA'}$ on the spin bundles if we demand

$$\nabla_a \epsilon_{BC} = 0 = \nabla_a \epsilon_{B'C'}. \qquad (4.1)$$

The discussion of spinor calculus parallels the review of tensor calculus in chapter 2.

The commutator of derivatives on a spinor will introduce curvature. We define the operator Δ_{AB} by

$$\nabla_a \nabla_b - \nabla_b \nabla_a = \epsilon_{A'B'} \Delta_{AB} + \epsilon_{AB} \Delta_{A'B'} \qquad (4.2)$$

so that

$$\Delta_{AB} = \nabla_{C'(A} \nabla_{B)}{}^{C'}. \qquad (4.3)$$

Note that this operator is linear over functions in that

$$\Delta_{AB}(f\alpha_C \bar{\beta}_{D'}) = f(\alpha_C \Delta_{AB}\bar{\beta}_{D'} + \bar{\beta}_{D'}\Delta_{AB}\alpha_C)$$

for any function f and pair of spinors $\alpha_C, \bar{\beta}_{D'}$. Thus we must have

$$\Delta_{AB}\alpha_C = \nabla_{E'(A}\nabla_{B)}{}^{E'}\alpha_C = \chi_{ABCD}\alpha^D \qquad (4.4)$$

$$\Delta_{A'B'}\alpha_C = \nabla_{E(A'}\nabla^E{}_{B')}\alpha_C = \Phi_{A'B'CD}\alpha^D \qquad (4.5)$$

for some curvature quantities χ_{ABCD} and $\Phi_{A'B'CD}$. From (4.1) and (4.3), these must have the symmetry properties

$$\chi_{ABCD} = \chi_{(AB)(CD)}$$
$$\Phi_{A'B'CD} = \Phi_{(A'B')(CD)}.$$

Further, from the Ricci identity of chapter 2 applied to an a.s.d. bivector

$$\Delta_{AB}\alpha^{(A}\beta^{B)} = 0$$

for any $\alpha^{(A}\beta^{B)}$. (This follows from the symmetry of the Ricci tensor which in turn follows from (2.2).) Thus

$$\chi_{AB}{}^B{}_D = 3\Lambda\epsilon_{AD}$$

for some Λ. We may therefore decompose χ_{ABCD} as

$$\chi_{ABCD} = \Psi_{ABCD} - 2\Lambda\epsilon_{D(A}\epsilon_{B)C} \tag{4.6}$$

where Ψ_{ABCD} is totally symmetric. To relate the three curvature quantities $\Psi_{ABCD}, \Phi_{A'B'CD}$ and Λ obtained in this way to the Riemann tensor, we have from (4.3), (4.4) and (4.5):

$$
\begin{aligned}
\Delta_{AB}(\alpha_C\bar{\beta}_{C'}) &= \bar{\beta}_{C'}\chi_{ABCD}\alpha^D + \alpha_C\bar{\Phi}_{ABC'D'}\bar{\beta}^{D'} \\
&= -(\chi_{ABCD}\epsilon_{C'D'} + \bar{\Phi}_{ABC'D'}\epsilon_{CD})\alpha^D\bar{\beta}^{D'} \\
&= -R_{ABCC'DD'}\alpha^D\bar{\beta}^{D'}
\end{aligned}
$$

where $R_{abcd} = R_{ABCC'DD'}\epsilon_{A'B'}$ + complex conjugate. Comparing this with the spinor decomposition of the Riemann tensor found in exercise 3d we find

$$
\begin{aligned}
C_{abcd} &= \Psi_{ABCD}\epsilon_{A'B'}\epsilon_{C'D'} + \text{complex conjugate} \tag{4.7} \\
R_{ab} - \frac{1}{4}Rg_{ab} &= -2\Phi_{ABA'B'} \tag{4.8} \\
R &= 24\Lambda \tag{4.9}
\end{aligned}
$$

for the Weyl tensor, trace-free Ricci tensor and Ricci scalar. In particular, this means that $\Phi_{ABA'B'}$ is Hermitian and Λ is real.

Note that the Einstein vacuum field equations are just

$$\Phi_{ABA'B'} = 0 = \Lambda.$$

The (second) Bianchi identity (2.3) translates to

$$\nabla_{A'}{}^A\Psi_{ABCD} = \nabla_{(B}{}^{B'}\Phi_{CD)A'B'}$$

$$\nabla^{AA'}\Phi_{ABA'B'} + 3\nabla_{BB'}\Lambda = 0.$$

In vacuum, this reduces to the vacuum Bianchi identity:

$$\nabla_{A'}{}^A\Psi_{ABCD} = 0. \tag{4.10}$$

Since the Weyl spinor Ψ_{ABCD} is totally symmetric, it will factorise

$$\Psi_{ABCD} = \alpha_{(A}\beta_B\gamma_C\delta_{D)}.$$

The factors define the four principal null directions (p.n.d.s) of the Weyl tensor (Pirani 1965; Penrose 1968b). The Petrov–Pirani–Penrose classification of Weyl tensors is based on a consideration of coincidences among

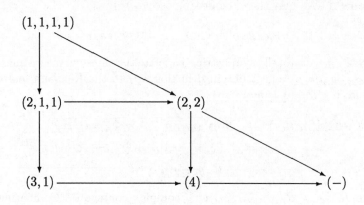

Figure 4.2. The PPP classification.

the p.n.d.s. These may all be distinct, when the Weyl tensor is said to be *algebraically general* or Type 1 (symbolically $(1,1,1,1,)$) or may coincide in various ways, when the Weyl tensor is said to be *algebraically special*. The possible coincidences may be represented as $(2,1,1)$, $(2,2)$, $(3,1)$ or (4), which are known respectively as Types 2, D, 3, and N, or the Weyl tensor may vanish $(-)$. The various cases may be arranged in a table of increasing specialisation as in figure 4.2.

We shall end this section with some examples of spinor field equations.

An electromagnetic field is represented by a bivector F_{ab}. The *source-free Maxwell equations* on F_{ab} are just

$$\nabla_{[a}F_{bc]} = 0; \ \nabla^a F_{ab} = 0.$$

If we introduce the a.s.d. part of F_{ab} :

$$W_{ab} = F_{ab} + i\,^*F_{ab}$$

then we can write these compactly as

$$\nabla^a W_{ab} = 0. \tag{4.11}$$

In spinor terms $W_{ab} = \varphi_{AB}\epsilon_{A'B'}$ for a symmetric spinor φ_{AB} and then (4.11) is just

$$\nabla_{A'}{}^A \varphi_{AB} = 0. \tag{4.12}$$

A linearised solution of Einstein's vacuum equations in Minkowski space is represented by a linearised Weyl spinor satisfying the vacuum Bianchi identity, i.e. by a symmetric spinor field φ_{ABCD} satisfying (4.10):

$$\nabla_{A'}{}^A \varphi_{ABCD} = 0. \tag{4.13}$$

Both (4.12) and (4.13) are examples of a general class of field equations – the *zero rest mass free field equations* or z.r.m. equations for brevity. The field is a symmetric spinor $\varphi_{AB...C}$ and the field equation is

$$\nabla_{A'}{}^A \varphi_{AB...C} = 0. \tag{4.14}$$

If the spinor has $2s$ indices, this represents a physical field of spin or *helicity* s. (Strictly speaking the field has positive helicity $+|s|$ if it has *positive frequency* and negative helicity $-|s|$ if it has *negative frequency*. Here positive frequency means that the Fourier transform of the field has only positive frequency components. We shall deal with this point more fully later on.) Thus the Maxwell field has spin 1 and the (linearised) gravitational field has spin 2. The spin $\frac{1}{2}$ member of the set is the Dirac or massless neutrino equation:

$$\nabla_{A'}{}^A \nu_A = 0. \tag{4.15}$$

The other type of equation which we shall be discussing in some detail is the *twistor equation*:

$$\nabla_{A'}{}^{(A} \omega^{B)} = 0 \tag{4.16}$$

and its higher valence generalisations:

$$\nabla_{A'}{}^{(A} \omega^{B...C)} = 0 \tag{4.17}$$

where $\omega^{B...C}$ is symmetric. For the moment we simply observe that, if (4.16) is satisfied then

$$\nabla_{A'}{}^A \omega^B = -i\epsilon^{AB} \pi_{A'} \tag{4.18}$$

for some other spinor field $\pi_{A'}$, so that (4.16) and (4.18) are equivalent.

Exercises 4

a) Consider the total space of one of the bundles H^k of exercise 3g as a real four-dimensional manifold. Show that if k is odd, then spinors cannot be defined on H^k.

b) A Maxwell field is *null* if the corresponding bivector F_{ab} is a null bivector. If the spinor representation of a null Maxwell field is $\varphi_{AB} = o_A o_B$ show that Maxwell's equation implies

$$o^A o^B \nabla_{AA'} o_B = 0$$

and that $\varphi_{AB} = f(x)\varphi_{AB}$ also defines a Maxwell field if

$$o^A \nabla_{AA'} f = 0.$$

What are the analogous statements for a linearised Type N Weyl spinor $\varphi_{ABCD} = o_A o_B o_C o_D$?

c) Show that, if ω_i^A are solutions of (4.16) for $i = 1, \ldots, n$ then $\omega_1^{(A} \ldots \omega_n^{B)}$ satifies (4.17).

d) If $\varphi_{A\ldots B}$ is a spin $2s$ solution of (4.14) and $(\omega^A, \pi_{A'})$ satisfies (4.18) show that

$$\chi_{A\ldots B} = \varphi_{A\ldots BC}\omega^C$$
$$\psi_{A\ldots C} = \bar{\omega}^{A'} \nabla_{AA'}\varphi_{B\ldots C} - i(s+1)\bar{\pi}_{(A}\varphi_{B\ldots C)}$$

are respectively spin $(2s - 1)$ and spin $(2s + 1)$ solutions of (4.14). These processes are referred to as *spin-lowering* and *spin-raising* (Penrose 1975; Hughston and Ward 1979).

e) If ω^{AB} is a solution of

$$\nabla^{A'(A}\omega^{BC)} = 0$$

so that $\nabla^{AA'}\omega^{BC} = -i\epsilon^{A(B}P^{C)A'}$ for some P^a, show that P^a is a Killing vector and further that $\nabla_a P_b$ is a self-dual bivector.

Chapter 5

Compactified Minkowski Space

A natural question to consider with spinor calculus is how to define the Lie derivative of a spinor α^A along a vector field X^a. If we had a definition satisfying the Leibniz property then necessarily the Lie derivative of the spinor ϵ_{AB}, $\mathcal{L}_X \epsilon_{AB}$, being skew would have to be a multiple of ϵ_{AB}

$$\mathcal{L}_X \epsilon_{AB} = \lambda \epsilon_{AB}.$$

However, this would entail

$$\mathcal{L}_X g_{ab} = (\lambda + \overline{\lambda}) g_{ab} = k g_{ab}. \tag{5.1}$$

Thus we could hope to define $\mathcal{L}_X \alpha^A$ only for vectors X^a satisfying (5.1). These are *conformal Killing vectors* and correspond to infinitesimal conformal isometries, that is maps P of M to itself with

$$P_*(g_{ab}) = k g_{ab}.$$

In flat space, **M**, these transformations define the *conformal group* $C(1,3)$ and we now turn to a consideration of this group.

By exercise 2c, and redefining k, equation (5.1) can be written

$$\nabla_a X_b + \nabla_b X_a = k g_{ab} \tag{5.2}$$

from which, by taking the trace,

$$k = \frac{1}{2} \nabla_a X^a. \tag{5.3}$$

We shall find the general solution of (5.2). Differentiating again we have

$$\nabla_a \nabla_b X_c + \nabla_a \nabla_c X_b = \nabla_a k g_{bc}.$$

Taking the cyclic sum $(abc) + (bca) - (cab)$ gives

$$\nabla_a \nabla_b X_c = \frac{1}{2}(\nabla_a k g_{bc} - \nabla_c k g_{ab} + \nabla_b k g_{ac}). \qquad (5.4)$$

Differentiating again gives

$$\nabla_a \nabla_b \nabla_c X_d = \frac{1}{2}(g_{cd}\nabla_a \nabla_b k - g_{bc}\nabla_a \nabla_d k + g_{bd}\nabla_a \nabla_c k).$$

Now since we are in flat space and there is no curvature, the skew part of this equation on (ab) must be zero:

$$g_{d[b}\nabla_{a]}\nabla_c k - g_{c[b}\nabla_{a]}\nabla_d k = 0.$$

From the trace on (cb):

$$2\nabla_a \nabla_d k + g_{ad}\Box k = 0$$

where $\Box k = g^{bc}\nabla_b \nabla_c k$. From the trace of this on (ad):

$$\begin{aligned}
\Box k &= 0 \\
\text{therefore } \nabla_a \nabla_b k &= 0, \\
\text{and so } k &= 2A + 4B_a x^a
\end{aligned}$$

for constant A and B_a. Thus by (5.4)

$$\nabla_a \nabla_b X_c = 2B_a g_{bc} - 2B_c g_{ab} + 2B_b g_{ac}.$$

Integrating once

$$\nabla_b X_c = 2B_a x^a g_{bc} - 2B_c x_b + 2B_b x_c + A g_{bc} + M_{bc}$$

for constant M_{bc}. Substituting into (5.2) gives $M_{bc} = -M_{cb}$. Integrating again finally gives

$$X_c = P_c + M_{bc}x^b + Ax_c + (2B_b x^b x_c - B_c x^b x_b) \qquad (5.5)$$

for constant P_c. The general conformal Killing vector therefore depends on 15 parameters. Ten of these, P_a and M_{ab}, correspond to the translations and rotations of the Poincaré group; A defines a *dilation* and the remaining four, B_a, define what are known as *special conformal transformations*. To understand these last we consider the integral curves of the vector field X^a obtained from (5.5) by setting P_a, M_{ab} and A to zero but retaining B_a. These integral curves are the solutions of

$$\frac{dx^a}{ds} = 2B_b x^b x^a - B^a x^b x_b.$$

By a straightforward integration we find

$$x^a(s) = \frac{x^a(0) - sB^a x^b(0)x_b(0)}{1 - 2sB_b x^b(0) + s^2 B_b B^b x_c(0)x^c(0)}. \tag{5.6}$$

From this we see at once that the vector field X^a is *incomplete* on Minkowski space; that is we shall reach infinite X^a at finite parameter values s where the denominator in (5.6) vanishes. To give a proper discussion of the conformal group, we need to extend Minkowski space by adding some points at infinity. The special conformal transformations will then interchange the points at infinity with some finite points of **M**. We shall describe the points at infinity in two different ways. For the first we consider the $O(2,4)$ null cone (Penrose and Rindler 1986). That is we consider a six-dimensional real manifold with a flat metric of signature -2. In coordinates T, V, W, X, Y, Z the metric is

$$ds^2 = (dT)^2 + (dV)^2 - (dW)^2 - (dX)^2 - (dY)^2 - (dZ)^2. \tag{5.7}$$

The null cone N is simply given by

$$(T)^2 + (V)^2 - (W)^2 - (X)^2 - (Y)^2 - (Z)^2 = 0 \tag{5.8}$$

and we wish to consider PN, the space of generators of N. (For a later application we remark that PN is a quadric Q in $\mathbf{RP}^5$.)

Clearly PN is a compact four-dimensional manifold. To investigate its topology, we may intersect N with a five-sphere S^5:

$$(T)^2 + (V)^2 + (W)^2 + (X)^2 + (Y)^2 + (Z)^2 = 2. \tag{5.9}$$

This will intersect each generator twice. Solving (5.8) and (5.9) simultaneously we see that

$$
\begin{aligned}
(T)^2 + (V)^2 &= 1 \\
(W)^2 + (X)^2 + (Y)^2 + (Z)^2 &= 1.
\end{aligned}
$$

The intersection of N and S^5 is therefore $S^1 \times S^3$ topologically. This is a double cover of PN; to obtain PN we must identify antipodal points of the S^1 and of the S^3 simultaneously. However this is again topologically $S^1 \times S^3$ (exercise!).

By the same considerations as in chapter 3, any two cuts of N have metrics which are conformally related, so that PN has a conformal metric.

The group $O(2,4)$ is the group preserving the quadratic form (5.7) and so maps PN to itself preserving the conformal metric. Since both I and $-I$ in $O(2,4)$ give the identity map on PN, $O(2,4)$ is a double cover of the conformal group of the conformal metric on PN. To make the connection with Minkowski space we observe that the map

$$X^a \to (X^0, \frac{1}{2}(1 - X^b X_b), -\frac{1}{2}(1 + X^b X_b), X^1, X^2, X^3) \qquad (5.10)$$

embeds Minkowsi space in the null cone N. In fact the image of Minkowski space is the intersection of N with the hyperplane:

$$V - W = 1. \qquad (5.11)$$

Substituting (5.11) into (5.7), we see that the embedding is actually isometric.

On any generator of N with $V - W \neq 0$, we can find a point satisfying (5.11) and hence a point of **M**. Thus **M** is identified with a subset of PN, and the conformal metric of **M** agrees with the conformal metric of PN. PN is therefore a compactification $\mathbf{M}^c$ of **M**, such that the conformal metric of **M** extends to the whole compactification. The points in $\mathbf{M}^c$ not in **M** correspond to the generators of PN with $V - W = 0$. This is the intersection of N with a null hyperplane through the origin. All such hyperplanes are equivalent under $O(2,4)$ so to see what these extra points represent, we consider the null hyperplane

$$V + W = 0.$$

From (5.10) we see that the points of **M** corresponding to generators of N which lie in this hyperplane are just the null cone of the origin. Thus $\mathbf{M}^c$ consists of **M** with an extra null cone added at infinity. Conversely, $\mathbf{M}^c$ is a compact manifold with conformal structure and if we remove the null cone of one point then what is left is conformally related to **M**.

We have also seen by this construction that the group $O(2,4)$ is 2-1 isomorphic to the conformal group $C(1,3)$ of Minkowski space, and that the null cones of points in $\mathbf{M}^c$ are represented by the intersection of N with hyperplanes through the origin. An element of $O(2,4)$ leaving invariant the plane $V - W = 0$ will leave infinity invariant in $\mathbf{M}^c$ and so will act only on **M**. This corresponds to a Poincaré transformation or dilation. A reflection in the plane $V = 0$ defines an inversion in $\mathbf{M}^c$ which interchanges the null cone at infinity and the null cone of the origin. To see this, we write Δ for $x^b x_b$ and square brackets for the homogeneous coordinates on **PN**. Then the transformation is

$$[x^0, \frac{1}{2}(1-\Delta), -\frac{1}{2}(1+\Delta), x^1, x^2, x^3]$$

$$\rightarrow \quad [x^0, -\frac{1}{2}(1-\Delta), -\frac{1}{2}(1+\Delta), x^1, x^2, x^3]$$

$$= \quad [\frac{x^0}{\Delta}, \frac{1}{2}(1-\frac{1}{\Delta}), -\frac{1}{2}(1+\frac{1}{\Delta}), \frac{x^1}{\Delta}, \frac{x^2}{\Delta}, \frac{x^3}{\Delta}]$$

$$= \quad [\hat{x}^0, \frac{1}{2}(1-\hat{\Delta}), -\frac{1}{2}(1+\hat{\Delta}), \hat{x}^1, \hat{x}^2, \hat{x}^3]$$

where $\hat{x}^a = \frac{x^a}{\Delta}$ and $\hat{\Delta} = \hat{x}^a \hat{x}_a$.

We may now interpret the special conformal transformations: they consist of the sequence invert-translate-invert

$$x^a \rightarrow \frac{x^a}{\Delta} \rightarrow \frac{x^a}{\Delta} - sB^a \rightarrow \frac{x^a - sB^a \Delta}{1 - 2s(B.x) + s^2 B^2 \Delta}$$

where the translation is by $-sB^a$.

There is another way to describe the points at infinity in $\mathbf{M}^c$ which we shall now consider. The metric of Minkowski space written in terms of spherical polar coordinates is

$$ds^2 = dt^2 - dr^2 - r^2(d\theta^2 + \sin^2\theta d\varphi^2).$$

If we introduce in- and out-going null coordinates by

$$v = \frac{1}{\sqrt{2}}(t+r), \quad u = \frac{1}{\sqrt{2}}(t-r)$$

this takes the form

$$ds^2 = 2dudv - \frac{1}{2}(v-u)^2(d\theta^2 + \sin^2\theta d\varphi^2)$$

where the range of coordinates is $-\infty < u < v < +\infty$. We aim to add on the boundary points at $u, v = \pm\infty$. To do this, set $u = \tan p$, $v = \tan q$ so

$$ds^2 = \frac{1}{2}\sec^2 p \sec^2 q(4dpdq + \sin^2(q-p)(d\theta^2 + \sin^2\theta d\varphi^2)).$$

The range of coordinates is now $-\frac{\pi}{2} < p < q < \frac{\pi}{2}$. The points at infinity correspond to the places where the first factor is infinite. If we define another metric $d\hat{s}^2$ by

$$d\hat{s}^2 = \Omega^2 ds^2; \quad \Omega^2 = 2\cos^2 p \cos^2 q$$

Figure 5.1. Minkowski space on the Einstein Static Universe, representing
$\mathbf{R} \times S^3$ by $\mathbf{R} \times S^1$.

so that $d\hat{s}^2 = 4dpdq + \sin^2(q - p)(d\theta^2 + \sin^2 \theta d\varphi^2)$ then this metric is
conformally related to the Minkowski metric where $\Omega \neq 0$ but is perfectly
finite on a larger manifold. To see what the larger manifold is, set

$$p = \frac{1}{2}(T - R), \ q = \frac{1}{2}(T + R)$$

to find

$$ds^2 = dT^2 - dR^2 - \sin^2 R(d\theta^2 + \sin^2 \theta d\psi^2).$$

This is the metric on the manifold $\mathbf{R} \times S^3$ corresponding to the *Einstein
Static Universe* (Penrose 1965b; Hawking and Ellis 1973). The region corresponding to the Minkowski space is indicated φ_i on figure 5.1. The light
cone of the point i^- with coordinates $T = -\pi, R = 0$ is the null surface
$\mathcal{I}^-$. This refocusses at the antipodal point i° at $T = 0, R = \pi$. The light
cone of i° is $\mathcal{I}^+$ refocussing at $i^+, T = \pi, R = 0$. These surfaces therefore
bound the compactified Minkowski space as a manifold with boundary. In
the vicinity of $i^-(i^+)$, the Minkowski space $\mathbf{M}$ has the character of an interior future (past) light cone, while in the vicinity of i°, $\mathbf{M}$ is the exterior

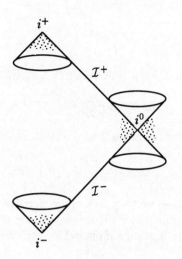

Figure 5.2. Compactified Minkowski space showing the neighbourhood of $\mathcal{I}$.

of the light cone as in figure 5.2. Therefore to make $\mathbf{M}^c$, the compactification of $\mathbf{M}$ discussed earlier, we identify opposite generators of $\mathcal{I}^-$ and $\mathcal{I}^+$, thereby identifying the points i^-, i° and i^+ as one point I. In this picture the $S^1 \times S^3$ topology is also manifest; a null geodesic is now topologically S^1 since its past endpoint on $\mathcal{I}^-$ is identified with its future endpoint on $\mathcal{I}^+$, while a space-like hypersurface has the topology of S^3, the one-point compactification (by $\mathcal{I}$) of $\mathbf{R}^3$.

Finally we remark the terms future/past *null infinity* for $\mathcal{I}^-/\mathcal{I}^+$ respectively since all null geodesics have a past endpoint on $\mathcal{I}^-$ and a future endpoint on $\mathcal{I}^+$; future/past *time-like infinity* for i^+/i^- since all time-like geodesics start from i^- and end at i^+; and *space-like infinity* for i° (Penrose 1965b).

We were led to the conformal group by a desire to define a Lie derivative for spinors. We may now give this definition. If X^a is any conformal Killing vector then the derivative of X^a must take the form

$$\nabla_a X_b = F_{ab} + \frac{k}{2} g_{ab} \qquad (5.12)$$

where $k = \frac{1}{2} \nabla_b X^b$ and F_{ab} is skew-symmetric so that

$$F_{ab} = \varphi_{AB} \epsilon_{A'B'} + \overline{\varphi}_{A'B'} \epsilon_{AB} \qquad (5.13)$$

for a symmetric spinor φ_{AB}. We define

$$\mathcal{L}_X \alpha^A = X^b \nabla_b \alpha^A - \varphi^A{}_B \alpha^B - \frac{k}{4} \alpha^A. \qquad (5.14)$$

We need to check that for a null vector $V^a = \alpha^A \beta^{A'}$ the Lie derivative satisfies

$$\mathcal{L}_X V^a = \alpha^A \mathcal{L}_X \overline{\beta}^{A'} + \overline{\beta}^{A'} \mathcal{L}_X \alpha^A$$

which is a straightforward calculation. From (5.14) and the Leibniz rule we deduce

$$\mathcal{L}_X \beta_A = X^b \nabla_b \beta_A + \varphi_A{}^B \beta_B + \frac{k}{4} \beta_A.$$

Given this definition we may seek those parts of spinor calculus which have a simple behaviour under Lie differentiation along conformal Killing vectors, as this will indicate a simple behaviour under conformal transformation.

The first example is the z.r.m. equations. For suppose $\psi_{A...B}$ is a solution of the spin s z.r.m. equation:

$$\nabla_{A'}{}^A \psi_{A...B} = 0$$

and consider

$$
\begin{aligned}
\chi_{A...B} &= \mathcal{L}_X \psi_{A...B} + \frac{1}{2} k \psi_{A...B} \qquad (5.15) \\
&= X^c \nabla_c \psi_{A...B} + 2s\varphi_{(A}{}^C \psi_{B...D)C} + \frac{s+1}{2} k \psi_{A...B}.
\end{aligned}
$$

From (5.13) and (5.5) we have

$$\nabla_{AA'} \varphi_{BC} = \frac{1}{2} \epsilon_{A(B} k_{C)A'} \qquad (5.16)$$

and from (5.15) and (5.16) it easily follows that

$$\nabla^{AA'} \chi_{A...B} = 0.$$

Thus the conformal group acts on solutions of the z.r.m. equations according to (5.15).

As another example, we consider the twistor equation. If

$$\beta^A = \mathcal{L}_X \alpha^A = X^E \nabla_E \alpha^A - \varphi^A{}_B \alpha^B - \frac{k}{2} \alpha^A \qquad (5.17)$$

then

$$\nabla^{A'(A}\beta^{B)} = \overline{\varphi}^{A'C'}\nabla^{(B}{}_{C'}\alpha^{A)} + \frac{k}{4}\nabla^{A'(A}\alpha^{B)} - \varphi^{B}{}_{C}\nabla^{A'(C}\alpha^{A)}$$
$$- \varphi^{A}{}_{C}\nabla^{A'(C}\alpha^{B)}$$

so that β^{A} will satisfy the twistor equation if α^{A} does.

Earlier in this chapter, we encountered the notion of *conformally rescaling* a metric g_{ab}. That is, we introduced a metric $\hat{g}_{ab}$ by

$$\hat{g}_{ab} = \Omega^{2}g_{ab} \tag{5.18}$$

While transformations from the conformal group may have this effect on g_{ab}, we are here concerned with something different. Conformal rescaling must change the connection: the Levi-Civita connection $\hat{\nabla}_{a}$ preserving $\hat{g}_{ab}$ evidently differs from ∇_{a} preserving g_{ab} by terms involving derivatives of Ω. Acting on a covector field V_{a} we find (Penrose and MacCallum 1973)

$$\hat{\nabla}_{a}V_{b} = \nabla_{a}V_{b} - \Upsilon_{a}V_{b} - \Upsilon_{b}V_{a} + g_{ab}g^{cd}\Upsilon_{c}V_{d} \tag{5.19}$$

where $\Upsilon_{a} = \nabla_{a}\ln\Omega$. For the effect on spinors, we define conformal rescaling by

$$\hat{\epsilon}_{AB} = \Omega\epsilon_{AB}; \ \hat{\epsilon}_{A'B'} = \Omega\epsilon_{A'B'}.$$

Then it is readily established that

$$\hat{\nabla}_{AA'}\alpha_{B} = \nabla_{AA'}\alpha_{B} - \Upsilon_{BA'}\alpha_{A} \tag{5.20}$$

so that also

$$\hat{\nabla}_{AA'}\beta^{B} = \nabla_{AA'}\beta^{B} + \delta_{A}{}^{B}\Upsilon_{CA'}\beta^{C}.$$

Now it is a sort of meta-theorem that Poincaré invariant theories with a simple or 'good' transformation under conformal rescaling have a simple or 'good' transformation under the conformal group and vice-versa. (There is a source of confusion in that both properties may be referred as 'conformal invariance'.)

As examples, we consider the effect of conformal rescaling on the z.r.m. equations and the twistor equation.

For the first we require that, under (5.18), a z.r.m. field $\varphi_{A...B}$ transforms as

$$\varphi_{A...B} \rightarrow \hat{\varphi}_{A...B} = \Omega^{-1}\varphi_{A...B}. \tag{5.21}$$

Then from (5.20) we deduce

$$\hat{\nabla}_{A'}{}^{A}\hat{\varphi}_{A...B} = \Omega^{-2}\nabla_{A'}{}^{A}\varphi_{A...B}$$

so that solutions of the z.r.m. equations go to solutions under conformal rescaling.

For the second, the correct transformation is

$$\hat{\omega}^A = \omega^A \tag{5.22}$$

so that

$$\hat{\nabla}_{AA'}\hat{\omega}^B = \nabla_{AA'}\omega^B + \delta_A{}^B \Upsilon_{CA'}\omega^C$$

and

$$\hat{\nabla}_{A'}{}^{(A}\hat{\omega}^{B)} = \Omega^{-1}\nabla_{A'}{}^{(A}\omega^{B)}.$$

We also deduce the transformation

$$\hat{\pi}_{A'} = \pi_{A'} + i\Upsilon_{AA'}\omega^A \tag{5.23}$$

where $\pi_{A'}$ is as in (4.18). From this it follows that the real scalar Σ defined by

$$\Sigma = \omega^A\bar{\pi}_A + \bar{\omega}^{A'}\pi_{A'} \tag{5.24}$$

is invariant under conformal rescaling. This result will be significant in chapter 7.

Exercises 5

a) There is an alternative way of defining $\mathcal{I}$ in terms of causal properties of **M**. Begin by defining a *causal curve* γ as one whose tangent is everwhere time-like or null and the *past* $J^-(S)$ of a set S as the set of all points $p \in \mathbf{M}$ which can be connected to S by future-pointing causal curves.

A causal curve when extended indefinitely into the future is held to define a point of *causal infinity*. Two curves define the same point iff they have the same pasts. Show that all time-like geodesics in **M** define a single point at infinity, i^+, in this sense. Show that two null geodesics in **M** define the same point at infinity iff they lie in a null hyperplane. Thus in this picture $\mathcal{I}^+$ appears as the set of all null hyperplanes. (For more on causal relations see e.g. Penrose 1972; Hawking and Ellis 1973.)

b) Define $P_{ab} = -\frac{1}{2}(R_{ab} - \frac{1}{6}Rg_{ab}) = \Phi_{ABA'B'} - \Lambda\epsilon_{AB}\epsilon_{A'B'}$. From the transformation (5.20) of the connection under conformal rescaling, deduce the transformation of the curvature:

$$\hat{\Psi}_{ABCD} = \Psi_{ABCD} \tag{5.25}$$

$$\hat{P}_{ab} = P_{ab} - \nabla_a\Upsilon_b + \Upsilon_a\Upsilon_b - \frac{1}{2}g_{ab}\Upsilon_c\Upsilon^c \tag{5.26}$$

where $\Upsilon_a = \nabla_a \ln\Omega$.

c) Show that under a special conformal transformation (5.6) the metric transforms like (5.18) with

$$\Omega = (1 - 2sB_b x^b(0) + s^2 B_b B^b x_c(0) x^c(0))^{-1}.$$

d) With Σ defined by (5.24) and $(\omega^A, \pi_{A'})$ satisfying the twistor equation, show directly that $\mathcal{L}_X \Sigma = 0$ for any conformal Killing vector X.

e) *De Sitter space* is a four-dimensional space of constant curvature arising as the hyperboloid

$$T^2 - W^2 - X^2 - Y^2 - Z^2 = -1$$

in the five-dimensional space with metric

$$ds^2 = dT^2 - dW^2 - dX^2 - dY^2 - dZ^2$$

(Hawking & Ellis 1973). By introducing coordinates

$$
\begin{aligned}
T &= \sinh t, \\
W &= \cosh t \sin \chi \sin \theta \sin \varphi, \\
X &= \cosh t \cos \chi, \\
Y &= \cosh t \sin \chi \cos \theta, \\
Z &= \cosh t \sin \chi \sin \theta \cos \varphi,
\end{aligned}
$$

show that de Sitter space is conformal to part of the Einstein Static Universe (Penrose 1965b). In particular this means that de Sitter space is conformally flat.

Chapter 6

The Geometry of Null Congruences

In this chapter, we discuss the geometry of null geodesic congruences in terms of spinors. We encounter the important notion of *shear* and its connection with complex analyticity which is one of the underlying themes of twistor theory.

A null geodesic congruence Γ in a region U of a space-time $\mathbf{M}$ is a set of null geodesics, one through each point of U. The tangent vectors to the geodesics of Γ define a field of null vectors l^a up to scale and so a spinor field o^A again up to scale. If we write D for $l^a \nabla_a$, the geodesic equation is

$$Dl^a = kl^a \text{ for some function } k$$

or equivalently

$$o^A Do_A = 0. \tag{6.1}$$

We may fix the scale of o_A along each geodesic by demanding

$$Do_A = 0 \tag{6.2}$$

which corresponds to parallel propagation of both the flag-pole and flag-plane of o_A.

To discuss the relative motion of nearby geodesics we consider *Jacobi fields*. A Jacobi field η^a along a particular null geodesic γ of Γ is simply a Lie-dragged vector:

$$\mathcal{L}_l \eta^a = 0 \text{ or } D\eta^a = \eta^b \nabla_b l^a. \tag{6.3}$$

45

From (6.3) and (6.2) we deduce the equation of *geodesic deviation*:

$$D^2\eta^a = R_{bcd}{}^a l^b \eta^c l^d \tag{6.4}$$

which relates the relative acceleration of geodesics neighbouring γ to the curvature tensor.

From (6.2) and (6.3) we see that $\eta^a l_a$ is constant along γ and so will always be zero if it is zero at one point of γ. The vanishing of $\eta^a l_a$ means that η^a lies in the null hypersurface Σ containing l^a so that the nearby geodesic represented by η^a is 'abreast' of γ (rather than earlier or later).

Next we introduce a spinor ι^A with

$$o_A \iota^A = 1; \ D\iota^A = 0. \tag{6.5}$$

This implies that the complex null vector $m^a = o^A \iota^{A'}$ is parallelly propagated along γ.

Now m^a and $\overline{m}^a$ span a two-plane element orthogonal to γ and lying in Σ; call this the 'screen-space', S. If η^a lies in S then

$$\eta^a = \overline{z}m^a + z\overline{m}^a, \ z \in \mathbf{C} \tag{6.6}$$

and we may think of S as the Argand plane of z. Substituting (6.6) into (6.3), contracting with m^a and changing the sign leads to

$$Dz = -\rho z - \sigma\overline{z} \tag{6.7}$$

where

$$\begin{aligned}
\rho &= m^a \overline{m}^b \nabla_b l_a = o^A \overline{m}^b \nabla_b o_A \\
\sigma &= m^a m^b \nabla_b l_a = o^A m^b \nabla_b o_A.
\end{aligned}$$

Thus the relative motion of neighbouring geodesics is described by the two complex functions ρ and σ. To disentangle their effects, we consider the progress of a small circle C in S, i.e. we follow a small circular tube of null geodesics near γ (Pirani 1965).

Suppose C is given initially by

$$z = \epsilon e^{i\varphi} \ ; \ 0 \le \varphi < 2\pi. \tag{6.8}$$

For a small affine parameter distance $\delta\lambda$ along γ, (6.7) gives

$$\delta z = -(\rho z + \sigma\overline{z})\delta\lambda. \tag{6.9}$$

We consider first the case

$$\sigma = 0, \ \rho = Re^{i\psi}.$$

(a)   **(b)**

Figure 6.1. The solid circle is changed (a) by convergence and rotation and (b) by shear.

Then the circle (6.8) becomes

$$z + \delta z = \epsilon e^{i\varphi}(1 - Re^{i\psi}\delta\lambda).$$

This represents another circle C'. The radius is diminished by the factor $(1 - R\cos\psi\delta\lambda)$ and the circle is rotated by an amount proportional to $R\sin\psi\delta\lambda$. We may write symbolically $\rho = $ convergence $+ i\times$rotation.

Now suppose

$$\rho = 0, \ \sigma = |\sigma|e^{2i\psi}$$

so that

$$z + \delta z = \epsilon(e^{i\varphi} - |\sigma|e^{i(2\psi-\varphi)}\delta\lambda).$$

To see what this represents, set

$$
\begin{aligned}
z + \delta z &= e^{i\psi}(x + iy) \\
x + iy &= \epsilon(e^{i(\varphi-\psi)}) - |\sigma|e^{i(\psi-\varphi)}\delta\lambda) \\
x &= \epsilon(1 - |\sigma|\delta\lambda)\cos(\varphi - \psi) \\
y &= \epsilon(1 + |\sigma|\delta\lambda)\sin(\varphi - \psi).
\end{aligned}
$$

This is an ellipse. The major and minor axes are $\epsilon(1 \pm |\sigma|\delta\lambda)$ and the inclination is $\psi = \frac{1}{2}\arg\sigma$. Note also that the area is unchanged to this order. The term for this behaviour is shear: σ is the *complex shear* (see figure 6.1).

This much of the discussion follows from (6.3). If we substitute (6.6) in (6.4) we find

$$
\begin{aligned}
D^2 z &= -R_{abcd}l^a m^b l^c \eta^d \\
&= -\Psi_0 \bar{z} - \Phi_{00} z
\end{aligned}
\tag{6.10}
$$

where

$$\Psi_0 = R_{abcd}l^a m^b l^c m^d = \Psi_{ABCD} o^A o^B o^C o^D \tag{6.11}$$

$$\Phi_{00} = \Phi_{ABA'B'} o^A o^B \bar{o}^{A'} \bar{o}^{B'} = -\frac{1}{2}R_{ab}l^a l^b. \tag{6.12}$$

From (6.7) and (6.10) we obtain the propagation equations for ρ and σ :

$$D\rho = \rho^2 + \sigma\bar{\sigma} + \Phi_{00} \tag{6.13}$$

$$D\sigma = (\rho + \bar{\rho})\sigma + \Psi_{00}. \tag{6.14}$$

These are known as the *Sachs equations* (Sachs 1961; Penrose 1968b) and allow an understanding of the effects of curvature on the congruence. From (6.13) we see that

$$D(\rho - \bar{\rho}) = (\rho - \bar{\rho})(\rho + \bar{\rho})$$

since, by (6.12), Φ_{00} is real.

This means that if there is no rotation at one point on γ, none will ever appear. We shall see in exercise 6b that the geometrical significance of no rotation is that the geodesics of the congruence lie in null hypersurfaces. From now on we shall suppose that ρ is real.

Next we remark that Φ_{00} is related by the Einstein equations to the energy-momentum tensor T_{ab} of the matter content of space-time:

$$\Phi_{00} = \frac{1}{2}T_{ab}l^a l^b.$$

For most 'normal' matter, Φ_{00} will be positive since it represents the local matter density measured by an observer whose world line is tangent to l^a. Of course, since l^a is null, no observer can have such a world line, but it is enough that the matter density be positive for a sequence of time-like vectors tending to l^a.

Thus if the geodesic γ passes through a region of Ricci curvature representing such normal matter, then by (6.13) ρ will increase, i.e. matter introduces convergence.

Finally, the effect of Ψ_0, which is a component of the Weyl tensor, is to introduce shear or, in the language of optics, *astigmatic* focussing. This in turn enters into (6.13) and causes convergence (Penrose 1966).

Returning to the screen space S, we observe that the relative motion of geodesics as recorded in S is *holomorphic* iff the shear vanishes. We may consider a related question: given a null hypersurface Σ ruled by null geodesics tangent to l^a, when is there a conformal metric on the space of generators? This metric would be of the form

$$\eta_{ab} = 2m_{(a}\overline{m}_{b)}$$

and to be a conformal metric on the space of generators its Lie derivative along the generators must be proportional to itself, modulo terms tangent to the generators. That is, we require

$$\mathcal{L}_l m_{(a} \overline{m}_{b)} = k m_{(a} \overline{m}_{b)} + l_{(a} V_{b)} \tag{6.15}$$

for some scalar k and vector V_a. (6.15) is equivalent to

$$m^a m^b \mathcal{L}_l m_{(a} \overline{m}_{b)} = 0$$

which, by the definition of σ, is just $\sigma = 0$.

Thus the (two-dimensional) space of generators of a null hypersurface Σ has a conformal metric iff the generators of Σ are shear-free. From the discussion in chapter 3, we know that a special case of such a Σ is the light cone of a point in Minkowski space.

Also from exercise 3h, for a two-dimensional manifold, having a conformal metric is equivalent to having a complex structure. Thus we have two related examples of the interplay between shear and complex analyticity.

To provide a third, we consider the problem of constructing shear-free congruences, that is, null geodesic congruences for which σ vanishes.

In conformally curved space, we see at once that this will be difficult. If σ is to vanish then from (6.14) so must Ψ_0 which means by (6.11) that o^A must be a principal null direction. At any point of space-time there are only four p.n.d.s so the possible shear-free congruences are few. In fact the situation is worse in that, by the Goldberg–Sachs theorem (Pirani 1965), a null geodesic congruence in vacuum is shear-free iff it is a *repeated* p.n.d. of the Weyl tensor. Thus, among vacuum space-times only the algebraically special ones admit shear-free congruences and they have just one each, except for Type D which has two.

We therefore confine our attention to flat space.

From (6.1) and (6.7) the conditions for a congruence to be geodesic and shear-free (g.s.f.) are

$$o^A Do_A \equiv o^A o^B \overline{o}^{B'} \nabla_{BB'} o_A = 0$$
$$\sigma \equiv o^A o^B \iota^{B'} \nabla_{BB'} o_A = 0.$$

These conditions may be combined to give the g.s.f. condition as

$$o^A o^B \nabla_{BB'} o_A = 0. \tag{6.16}$$

We remark that the condition in this form is independent of the scaling of o^A, and also that it is holomorphic in the components of o^A, i.e. there are no appearances of $\bar{o}^{A'}$. To solve (6.16) we choose a constant normalised dyad (α_A, β_A) and coordinatise o^A as

$$o^A = \lambda(\alpha^A - L\beta^A) \text{ so that } L = \frac{o_0}{o_1} = -\frac{o^1}{o^0}. \tag{6.17}$$

Then (6.16) reduces to the two equations

$$\nabla_{0A'}L - L\nabla_{1A'}L = 0; \quad A' = 0', 1'. \tag{6.18}$$

We write these equations out in Minkowski coordinates

$$x^{AA'} = \begin{bmatrix} u & \zeta \\ \bar{\zeta} & v \end{bmatrix} = \frac{1}{\sqrt{2}} \begin{bmatrix} t+z & x+iy \\ x-iy & t-z \end{bmatrix} \tag{6.19}$$

as

$$\frac{\partial L}{\partial u} - L\frac{\partial L}{\partial \bar{\zeta}} = 0; \quad \frac{\partial L}{\partial \zeta} - L\frac{\partial L}{\partial v} = 0.$$

By the method of characteristics, the general analytic solution of these equations is given implicitly in terms of an arbitrary analytic function of three variables:

$$F(Lu + \bar{\zeta}, L\zeta + v, L) = 0. \tag{6.20}$$

This can be written more symmetrically, taking account of (6.17), in terms of an arbitrary analytic function of four variables which is homogeneous of some degree

$$f(\lambda W_0, \lambda W_1, \lambda W_2, \lambda W_3) = \lambda^k f(W_0, W_1, W_2, W_3)$$

as

$$f(-ix^{A'B}o_B, o_B) = 0 \tag{6.21}$$

where

$$F(a, b, c) = f(-ia, -ib, c, 1).$$

So the general analytic g.s.f. congruence in flat space is defined by an arbitrary homogeneous holomorphic function of four variables. We shall find a geometrical interpretation of this result, which is known as the Kerr Theorem (Penrose 1967), in the next section.

Exercises 6

a) A vector field V^a is said to be *hypersurface-orthogonal* (h.s.o.) if it is orthogonal to a family of hypersurfaces, or equivalently if it is proportional to a gradient. Show that a necessary and sufficient condition for this is

$$V_{[a}\nabla_b V_{c]} = 0.$$

b) Show that the null vector field l^a is h.s.o. iff l^a is geodesic and $\rho = \bar{\rho}$.

c) If t^a is the tangent vector to the world-line γ of the origin of coordinates in Minkowski space, show that the g.s.f. congruence obtained from the function

$$f = t_{A'}{}^A x^{A'B} o_A o_B$$

consists of the generators of the in- and out-going null cones springing from γ.

Chapter 7

The Geometry of Twistor Space

Our first definition of a twistor is as a spinor field $\Omega^A(x)$ in Minkowski space $\mathbf{M}$ satisfying the twistor equation

$$\nabla_{A'}{}^{(A}\Omega^{B)} = 0 \tag{7.1}$$

or equivalently

$$\nabla_{AA'}\Omega^B = -i\delta_A{}^B\pi_{A'} \tag{7.2}$$

for some other spinor field $\pi_{A'}$.

Our first task is therefore to solve this equation. Since there is no curvature we have

$$\nabla_{A(A'}\nabla_{B')}{}^A\Omega_B = 0 = \nabla_{A'(A}\nabla^{A'}{}_{B)}\Omega_C.$$

Substituting from (7.2) we have at once that

$$\nabla_{AA'}\pi_{B'} = 0$$

and so (7.2) can be integrated immediately:

$$\Omega^A = \omega^A - ix^{AA'}\pi_{A'} \tag{7.3}$$

where the constant spinor ω^A is a constant of integration.

Thus *twistor space* $\mathbf{T}$, the vector space of solutions to (7.1), is a four-dimensional complex vector space and may be coordinatised with respect to a choice of origin, by a pair of spinors,

$$(\omega^A, \pi_{A'}) = Z^\alpha \; ; \; \alpha = 0, 1, 2, 3. \tag{7.4}$$

We have seen already that the conformal group acts linearly on solutions of (7.1) via

$$\mathcal{L}_X \Omega^A = -iX^{AB'}\pi_{B'} - \varphi^A{}_B \Omega^B - \frac{1}{4}k\Omega^A$$

$$\mathcal{L}_X \pi_{A'} = \overline{\varphi}_{A'}{}^{B'}\pi_{B'} + \frac{1}{4}k\pi_{A'}$$

so that also

$$\mathcal{L}_X (\Omega^A \overline{\pi}_A + \overline{\Omega}^{A'}\pi_{A'}) = 0.$$

We may now interpret this as saying that the conformal group acts linearly on **T** and preserves the pseudo-Hermitian inner product

$$\Sigma = \Omega^A \overline{\pi}_A + \overline{\Omega}^{A'}\pi_{A'} = \omega^A \overline{\pi}_A + \overline{\omega}^{A'}\pi_{A'} \tag{7.5}$$

where we have used (7.3). Furthermore, no conformal transformation leaves all of twistor space invariant. (7.5) may be written

$$\Sigma = Z^0 \overline{Z}^2 + Z^1 \overline{Z}^3 + \overline{Z}^0 Z^2 + \overline{Z}^1 Z^3 \tag{7.6}$$
$$= \Sigma_{\alpha\beta'} Z^\alpha \overline{Z}^{\beta'}$$

where $\overline{Z}^{\beta'} \in \overline{\mathbf{T}}$ is the complex conjugate of $Z^\beta \in \mathbf{T}$. Since the inner product Σ is non-degenerate, we may use it to identify $\overline{\mathbf{T}}$ with $\mathbf{T}^*$ so that primed twistor indices never appear. We therefore view complex conjugation as a map to $\mathbf{T}^*$ rather than $\overline{\mathbf{T}}$:

$$Z^\alpha = (\omega^A, \pi_{A'}) \to \overline{Z}_\alpha = (\overline{\pi}_A, \overline{\omega}^{A'}).$$

From the ϵ_{AB} and $\epsilon_{A'B'}$ on Minkowski space, we define a totally skew twistor $\epsilon_{\alpha\beta\gamma\delta}$ by

$$\epsilon_{\alpha\beta\gamma\delta} Z^\alpha_{\;i} Z^\beta_{\;j} Z^\gamma_{\;k} Z^\delta_{\;l} = \epsilon_{AB} \omega^A_{[i} \omega^B_{j} \epsilon^{A'B'} \pi_{A'}{}_{k} \pi_{B'}{}_{l]} \tag{7.7}$$

It is easy to see that this is also invariant under the conformal group.

The signature of Σ is zero, so that the endomorphisms of **T** preserving Σ constitute the group $U(2,2)$. If we restrict to endomorphisms preserving $\epsilon_{\alpha\beta\gamma\delta}$, this is reduced to $SU(2,2)$ which has fifteen parameters. Above we saw that $C(1,3)$ acts linearly on **T** preserving Σ. Since $C(1,3)$ also has fifteen parameters these groups must be locally isomorphic.

We shall see their relation more precisely below. This will give us a second definition of twistor space as the representation space of the spinor representation for $SU(2,2)$.

For a third definition, we consider for a particular field (7.3) the points of (complexified) Minkowksi space at which the field vanishes. These satisfy

$$\omega^A = ix^{AA'}\pi_{A'}. \tag{7.8}$$

Evidently, we must allow x^a to be complex to find solutions of this in general. By complexified Minkowski space, written $\mathbf{CM}$, we mean just $\mathbf{C}^4$ with the usual Minkowski metric:

$$(z^0)^2 - (z^1)^2 - (z^2)^2 - (z^3)^2$$

(in particular, *not* a Hermitian metric).

The general solution of (7.8) is given in terms of a particular solution $x_0^{AA'}$ by

$$x^{AA'} = x_0^{AA'} + \lambda^A \pi^{A'} \tag{7.9}$$

where λ^A is an arbitrary spinor.

This defines a 2-plane in $\mathbf{CM}$, every tangent of which has the form $\lambda^A \pi^{A'}$ for varying λ^A. Thus every tangent is null and any two are orthogonal. Further the tangent bivector is self-dual. Such a *totally null* 2-plane is referred to as an *α-plane*. (A *β-plane* is a totally null 2-plane with anti-self-dual bivector and arises as the zeroes of a solution of the complex conjugate twistor equation, i.e. from a *dual* twistor.)

Note that the α-plane itself is defined by the proportionality class $[Z^\alpha]$ of the twistor Z^α. If we define the *projective twistor space* $\mathbf{PT}$ as the projective version of $\mathbf{T}$, then the points of $\mathbf{PT}$ correspond to α-planes. Likewise the points of $\mathbf{PT^*}$ correspond to β-planes. The extra information in $\mathbf{T}$ is the choice of scale for the spinor $\pi_{A'}$ associated to a particular α-plane. $\mathbf{PT}$ and $\mathbf{PT^*}$ are examples of $\mathbf{CP}^3$ (see exercise 3e).

In general, an α-plane will have no real point. If one does, say $x_0^{AA'}$ in (7.9) is real, then from (7.8)

$$\omega^A \overline{\pi}_A = ix_0^{AA'}\overline{\pi}_A \pi_{A'}$$

so that from (7.5), Σ vanishes. The corresponding twistor is said to be *null*. Conversely, if Σ is zero then

$$\omega^A \overline{\pi}_A = ia, \ a \in \mathbf{R}.$$

Define

$$x_0^{AA'} = \frac{1}{a}\omega^A \overline{\omega}^{A'}$$

then

$$\omega^A = ix_0^{AA'}\pi_{A'}$$

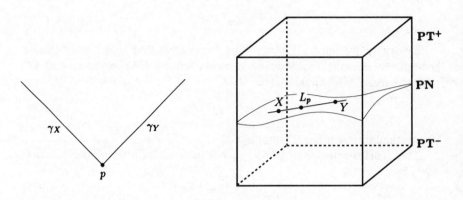

Figure 7.1. The null geodesics γ_X and γ_Y meeting at p correspond to null twistors X and Y lying on the projective line L_p.

and $x_0^{AA'}$ is a real point on the α-plane. (If a happens to be zero a slightly different construction of a real point on the α-plane is necessary.) Thus an α-plane contains a real point iff the corresponding twistor is null. The α-plane will then contain the whole null geodesic:

$$x^{AA'} = x_0^{AA'} + r\overline{\pi}^A\pi^{A'}, \ r \in \mathbf{R}.$$

We may divide $\mathbf{T}$ and $\mathbf{PT}$ into three parts respectively $\mathbf{T}^+, \mathbf{T}^-$ and N or $\mathbf{PT}^+, \mathbf{PT}^-$ and $\mathbf{PN}$ according as Σ is positive, negative or zero. Then we have seen that $\mathbf{PN}$ corresponds to unscaled real null geodesics in $\mathbf{M}$. We proceed to extend this correspondence between geometry in $\mathbf{M}$ or $\mathbf{CM}$ and geometry in $\mathbf{T}$ or $\mathbf{PT}$ (figure 7.1).

First we ask: given two null twistors X^α, Y^α representing null geodesics in $\mathbf{M}$, what is the condition for the null geodesics to meet?

We have

$$X^\alpha = (ix_0^{AA'}X_{A'}, X_{A'}); \ Y^\alpha = (ix_1^{AA'}Y_{A'}, Y_{A'})$$

with corresponding null geodesics

$$\gamma_X : x^{AA'} = x_0^{AA'} + \lambda\overline{X}^A X^{A'}; \ \gamma_Y : x^{AA'} = x_1^{AA'} + \mu\overline{Y}^A Y^{A'}.$$

If these intersect at say $x_2^{AA'}$ then

$$x_2^{AA'} = x_0^{AA'} + \lambda\overline{X}^A X^{A'} = x_1^{AA'} + \mu\overline{Y}^A Y^{A'} \qquad (7.10)$$

for some real λ and μ, so that

$$x_2^{AA'}\overline{Y}_A X_{A'} = x_0^{AA'}\overline{Y}_A X_{A'} = x_1^{AA'}\overline{Y}_A X_{A'}$$

and

$$X^\alpha \overline{Y}_\alpha = i(x_0^{AA'}\overline{Y}_A X_{A'} - x_1^{AA'}\overline{Y}_A X_{A'}) = 0. \tag{7.11}$$

Conversely, given (7.11) we deduce that

$$x_0^{AA'} - x_1^{AA'} = -\lambda \overline{X}^A X^{A'} + \mu \overline{Y}^A Y^{A'}$$

for some real λ and μ, provided $X_{A'}Y^{A'} \neq 0$. If $X_{A'}Y^{A'}$ is zero, so that γ_X and γ_Y are parallel, then from (7.11) we deduce instead that

$$x_0^{AA'} - x_1^{AA'} = \lambda \overline{X}^A X^{A'} + \zeta^A X^{A'} + \overline{\zeta}^{A'} \overline{X}^A$$

for some ζ^A and real λ. If ζ^A is zero, then γ_X and γ_Y coincide, while if ζ^A is non-zero then γ_X and γ_Y are two generators of a null hypersurface and consequently they intersect at $\mathcal{I}$.

Thus, in all cases, the null geodesics γ_X and γ_Y will intersect iff $X^\alpha \overline{Y}_\alpha$ is zero.

Since also $X^\alpha \overline{X}_\alpha$ and $Y^\alpha \overline{Y}_\alpha$ are zero this means that any linear combination

$$Z^\alpha = \zeta X^\alpha + \eta Y^\alpha; \ \zeta, \eta \in \mathbf{C}$$

will be null and will define a null geodesic which meets the previous two.

All these geodesics therefore define the null cone of a point in $\mathbf{M}$: the null cone of a point in $\mathbf{M}$ is a two-dimensional subspace of $\mathbf{T}$ lying entirely in N or a projective line $L_p = \mathbf{P}^1$ in $\mathbf{PT}$ lying in $\mathbf{PN}$. The equation of L_p is again

$$\omega^A = i x^{AA'} \pi_{A'}$$

where we now think of $x^{AA'}$ as fixed (the coordinate of p) and $(\omega^A, \pi_{A'})$ as varying.

The points of $\mathbf{PT}$ lying on L_p correspond to the various projective spinors at p, i.e. to the projective null cone or celestial sphere of p.

In $\mathbf{M}$, two points p and q are null separated iff there is a null geodesic connecting them. Translated to $\mathbf{PT}$, this means iff the lines L_p and L_q intersect, since the point of intersection represents the connecting null geodesic.

The points of compactified Minkowski space $\mathbf{M}^c$ not included in $\mathbf{M}$ make up the null cone of the point at infinity, I. There is therefore a special line in $\mathbf{PN}$, also written I, and $\mathcal{I}$ is represented by all the lines which intersect this one. In coordinates $(\omega^A, \pi_{A'})$, the line I has equation $\pi_{A'} = 0$.

We now consider the lines in **PT** which don't lie entirely in **PN**. An arbitrary line may be represented in terms of two points X^α, Y^α on it by the bivector

$$P^{\alpha\beta} = X^\alpha Y^\beta - Y^\alpha X^\beta = 2X^{[\alpha} Y^{\beta]}. \tag{7.12}$$

Evidently the proportionality class of $P^{\alpha\beta}$ depends only on the line and not on the choice of points on the line. In the space of proportionality classes of all possible bivectors, which is a projective space $\mathbf{P}^5$, the lines of **PT** are the simple ones. By exercise 2b, these are just the solutions of the equation

$$\epsilon_{\alpha\beta\gamma\delta} P^{\alpha\beta} P^{\gamma\delta} = 0. \tag{7.13}$$

(7.13) is the equation of a quadric in $\mathbf{P}^5$. Thus the space of lines in **PT** is represented by a compact complex manifold Q_4 in $\mathbf{P}^5$; this is known as the *Klein representation*. (In the language of exercise 3e, Q_4 is the Grassmannian $G(2, 4)$ of two planes in the four-dimensional complex vector space **T**.) It also follows from exercise 2b that if $P^{\alpha\beta}$ and $Q^{\alpha\beta}$ are both simple then the corresponding lines in **PT** will intersect iff

$$\epsilon_{\alpha\beta\gamma\delta} P^{\alpha\beta} Q^{\gamma\delta} = 0. \tag{7.14}$$

We define a conformal metric on Q_4 by saying that two points are null separated if the corresponding lines intersect in **PT**. To see that this is a quadratic condition, consider a particular point $P^{\alpha\beta}$ on Q_4. Then (7.14) for all possible $Q^{\alpha\beta}$ defines the tangent plane to Q_4 at $P^{\alpha\beta}$. The intersection of this tangent plane with Q_4 is therefore a cone, and this is the null cone of $P^{\alpha\beta}$.

With this conformal metric, Q_4 is $\mathbf{CM}^c$, that is *complexified compactified* Minkowski space.

To locate the real points we need to use the Hermitian structure on **PT**. We recall that a line L defined a bivector $P^{\alpha\beta}$ as the outer product of any two points on it. We could equally have used the bivector $P_{\alpha\beta}$ defined as the outer product of any two planes through L. This is the dual description

$$P_{\alpha\beta} = \frac{1}{2}\epsilon_{\alpha\beta\gamma\delta} P^{\gamma\delta}.$$

Also, given a line $P^{\alpha\beta}$, we can define a complex conjugate line $\overline{P}_{\alpha\beta}$ using the Hermitian structure. We may therefore define a line as real if

$$\overline{P}_{\alpha\beta} = \frac{1}{2}\epsilon_{\alpha\beta\gamma\delta} P^{\gamma\delta}. \tag{7.15}$$

It is then an exercise to show that the line $P^{\alpha\beta}$ is real iff all the points Z^α on $P^{\alpha\beta}$ are null.

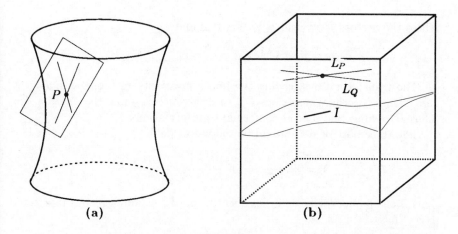

Figure 7.2. The tangent plane at P to Q_4, (a), is represented in **PT**, (b), by the lines L_Q intersecting L_P. Infinity is represented by the line I in **PN**.

The space of projective bivectors satisfying (7.15) is a *real* projective space $\mathbf{RP}^5$ and the simple ones give a *real* quadric $\mathbf{R}Q_4$. This is the real compactified Minkowski space $\mathbf{M}^c$ of chapter 5, and in suitable coordinates, (7.13) is (5.8).

The group $SU(2,2)$ acts on **PT** and so on $\mathbf{R}Q_4$. Since both groups have fifteen parameters, this gives a local isomorphism of $SU(2,2)$ and $O(2,4)$. However, minus the identity in $SU(2,2)$ gives the identity in $O(2,4)$. Also, as we have seen, minus the identity in $O(2,4)$ gives the identity in $C(1,3)$, so we have a chain of isomorphisms

$$SU(2,2) \overset{2-1}{\to} O(2,4) \overset{2-1}{\to} C(1,3).$$

The infinitesimal generators of $SU(2,2)$ may be written as vector fields

$$iE^\alpha{}_\beta Z^\beta \frac{\partial}{\partial Z^\alpha}$$

on **T**, where the matrices $E^\alpha{}_\beta$ are Hermitian and trace-free:

$$E^\alpha{}_\beta = \overline{E}_\beta{}^\alpha,\ E^\alpha{}_\alpha = 0.$$

In coordinates,

$$E^\alpha{}_\beta Z^\beta \frac{\partial}{\partial Z^\alpha} \;=\; E^A{}_B \omega^B \frac{\partial}{\partial \omega^A} + P^{AB'}\pi_{B'}\frac{\partial}{\partial \omega^A} + B_{A'B}\omega^B \frac{\partial}{\partial \pi_{A'}}$$

$$+ \overline{E}_{A'}{}^{B'} \pi_{B'} \frac{\partial}{\partial \pi_{A'}}$$

where the vectors $P^{AA'}$ and $B^{AA'}$ are real and

$$E_{AB} = \varphi_{AB} + A\epsilon_{AB}, \ \varphi_{AB} = \varphi_{(AB)}, \ A \in \mathbf{R}.$$

The transformations leaving the line I invariant are those with $B^{AA'}$ zero. These are the translations P^a, the rotations φ_{AB} and the dilation A. Thus B^a defines the special conformal transformations.

The trace part of $E^\alpha{}_\beta$ would be the vector field

$$\Upsilon = Z^\alpha \frac{\partial}{\partial Z^\alpha} = \omega^A \frac{\partial}{\partial \omega^A} + \pi_{A'} \frac{\partial}{\partial \pi_{A'}}. \tag{7.16}$$

This is the homogeneity or Euler operator on $\mathbf{T}$. A homogeneous function $f(Z^\alpha)$ has

$$\Upsilon f = nf$$

where n is the homogeneity. (Such an f should be thought of as a section of the bundle $\mathcal{O}(n)$ discussed in exercise 3g.)

In particular, a geometrical object w on $\mathbf{T}$ is defined on $\mathbf{PT}$ if it is Lie-dragged along Υ:

$$\mathcal{L}_\Upsilon w = 0. \tag{7.17}$$

In practice, one can tell by inspection whether (7.17) holds, since the requirement is just that w have homogeneity zero.

We conclude this chapter with a discussion of the *Kerr Theorem* which leads to a fourth picture of what a twistor is.

The zero set of an analytic function $f(Z^\alpha)$ of the twistor variable is well defined on $\mathbf{PT}$ provided f is homogeneous of some degree. Locally the zero set is a hypersurface which will intersect $\mathbf{PN}$ in a three-dimensional set K. Then K defines a congruence of null geodesics in $\mathbf{M}$. The content of the Kerr Theorem is that this congruence is shear-free and conversely that all analytic g.s.f. congruences arise in this way. To see this, recall that a point on N is a twistor $Z^\alpha = (\omega^A, \pi_{A'})$ with $\omega^A = ix^{AA'}\pi_{A'}$ for some real point $x^{AA'}$ and that Z^α defines the null geodesic γ through $x^{AA'}$ in the direction of $\overline{\pi}_A \pi_{A'}$. Thus given $f(Z^\alpha)$, we define K by

$$f(ix^{AA'}\pi_{A'}, \pi_{A'}) = 0$$

to obtain $\pi_{A'}$ as a function of $x^{AA'}$.

As an example, the simplest case is to take for f a linear function

$$f = A_\alpha Z^\alpha. \tag{7.18}$$

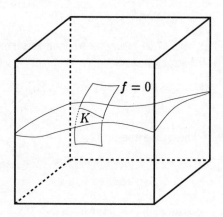

Figure 7.3. The Kerr theorem.

The resulting congruence, known as a *Robinson congruence* (Penrose 1967), will then give a picture of the (dual) twistor A_α. The calculation is straightforward, though a little messy. If $A_\alpha = (A_A, A^{A'})$ we have from (7.18)

$$(iA_A x^{AA'} + A^{A'})\pi_{A'} = 0$$

so that $\pi^{A'} = k(iA_A x^{AA'} + A^{A'})$. Choosing coordinates as in (6.19), suppose also

$$A_A = o_A; \quad A^{A'} = -a\bar{\iota}^{A'}$$

then $\pi^{A'} = ik(x - iy, ia + t - z)$. For convenience of visualisation, we choose a surface Σ of constant t and project the null vector $l^a = \bar{\pi}^A \pi^{A'}$ into Σ:

$$\hat{l}^a = l^a - t^a t^b l_b.$$

Now we seek the integral curves of $\hat{l}^a$ in Σ. This requires the solution of

$$\frac{dx^a}{ds} = \hat{l}^a$$

i.e.

$$\frac{d\zeta}{ds} = 2k\zeta(iat - z) \qquad (7.19)$$

$$\frac{dz}{ds} = k(\zeta\bar{\zeta} - a^2 - (t-z)^2) \qquad (7.20)$$

where $\zeta = x + iy$. If we choose the parameter s to be proper distance then

$$k = \zeta\bar{\zeta} + (t - z)^2 + a^2. \tag{7.21}$$

The solution of (7.19) with (7.21) may be written in terms of the arbitrary complex constant of integration α as

$$\begin{aligned} Q_1 &\equiv k + \alpha\zeta + \overline{\alpha\zeta} - 2a^2 = 0 \\ Q_2 &\equiv \alpha\zeta - \overline{\alpha\zeta} - 2ia(t - z) = 0. \end{aligned}$$

If we set $\alpha = \tan\theta e^{i\varphi}$ these become

$$Q_1 \equiv x^2 + y^2 + (t - z)^2 + 2ax\tan\theta\cos\varphi - 2ay\tan\theta\sin\varphi - a^2 = 0 \tag{7.22}$$

$$\frac{i}{2a}Q_2 \equiv (t - z) - x\tan\theta\sin\varphi - y\tan\theta\cos\varphi = 0. \tag{7.23}$$

If we eliminate φ and set $\rho^2 = x^2 + y^2$ we obtain

$$(\rho - a\sec\theta)^2 + (t - z)^2 = a^2\tan^2\theta. \tag{7.24}$$

Now (7.22) defines a sphere with centre $(-a\tan\theta\cos\varphi, a\tan\theta\sin\varphi, t)$ and (7.23) defines a plane through $(0, 0, t)$. Thus the integral curves are circles. These circles, for varying φ but constant θ, lie on (7.24) which defines a torus. These are not the 'obvious' circles obtained by intersecting the torus with horizontal and vertical planes. Instead they twist round the torus and each links all the others. The sense of the twisting depends on the sign of

$$a = -\frac{1}{2}A_\alpha\bar{A}^\alpha.$$

For varying θ, (7.24) defines a family of coaxial tori obtained by rotating a set of coaxial circles round the z-axis. The degenerate torus $\theta = 0$ is just the circle $\rho = a, t = z$. All the circles link through this degenerate torus.

The Robinson congruence is reconstructed by attaching torches tangent to all the circles on all the tori. These are switched on and the whole system is moved up the z-axis at the speed of light!

In the limit of zero a, i.e. when A_α is a null vector, the congruence degenerates in the spatial picture to the set of all radial rays from the point $x = y = t - z = 0$. Thus, in space-time, the congruence is defined by all the null geodesics meeting the particular null geodesic defined by A_α. For a non-null twistor, the geodesics just fail to meet. Instead they twist around the degenerate torus, which moves along at the speed of light.

Exercises 7

a) Verify that the signature of Σ in (7.7) is zero.

b) If α^A, β^A are two solutions of (7.1), show that $\alpha^A \bar{\beta}^{A'}$ is a conformal Killing vector.

c) Show that the line L corresponding to the point $z^a = x^a - iy^a$ in **CM** lies entirely in $\mathbf{PT}^+$ iff y^a is time-like and future pointing. The corresponding part of **CM**, written $\mathbf{CM}^+$, is known as the *future tube*.

d) A line L in **PT** is represented by a simple bivector $P^{\alpha\beta}$. Show that X^α lies on L iff $X^{[\alpha} P^{\beta\gamma]} = 0$.

e) Verify that the simple bivector $P^{\alpha\beta}$ is real in the sense of (7.15) iff all the points Z^α on the corresponding line are null.

f) There is another way of describing the relationship between projective twistor space **PT** and compactified, complexified Minkowski space $\mathbf{CM}^c$ which will be of use at the beginning of chapter 10. Regarding **T** as a four-dimensional complex vector space define

$$
\begin{aligned}
F_1 &= \{L_1 : L_1 \text{ is a one-dimensional subspace of } \mathbf{T}\} \\
F_2 &= \{L_2 : L_2 \text{ is a two-dimensional subspace of } \mathbf{T}\} \\
F = F_{1,2} &= \{(L_1, L_2) : L_1 \text{ and } L_2 \text{ as above with } L_1 \text{ a subspace of } L_2\}.
\end{aligned}
$$

Now define projections μ and ν to obtain the *double fibration*

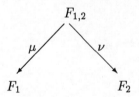

Show that F_1 can be identified with **PT**, F_2 with $\mathbf{CM}^c$, and $F_{1,2}$ with the primed spin bundle of $\mathbf{CM}^c$. Show also that

$$\mu(\nu^{-1}(\mathbf{CM})) = \mathbf{PT} - I.$$

g) Show that two α-planes always intersect but that the α-plane defined by A^α meets the β-plane defined by B_α iff $A^\alpha B_\alpha = 0$. In this case show that the intersection is a whole null geodesic.

h) Since de Sitter space is conformally flat, it will also be represented by part of the quadric Q_4 with the same definition of the conformal metric. To fix the conformal scale, pick a point $I^{\alpha\beta}$ of $\mathbf{RP}^5$ not on the quadric.

The polar plane of this point is represented by $I_{\alpha\beta} = \frac{1}{2}\epsilon_{\alpha\beta\gamma\delta}I^{\gamma\delta}$. Now for any two points $X^{\alpha\beta}, Y^{\alpha\beta}$ on the quadric consider the two-plane defined by $X^{\alpha\beta}, Y^{\alpha\beta}$ and $I^{\alpha\beta}$. This meets Q_4 in a projective line which meets $I_{\alpha\beta}$ in two more points, say $P^{\alpha\beta}$ and $Q^{\alpha\beta}$. Now define the distance d from $X^{\alpha\beta}$ to $Y^{\alpha\beta}$ by

$$d = \frac{1}{2}|\log|(x, y; p, q)||$$

where $(x, y; p, q)$ is the cross-ratio of $X^{\alpha\beta}, Y^{\alpha\beta}, P^{\alpha\beta}$ and $Q^{\alpha\beta}$. Show that this agrees with the definition of de Sitter space in exercise 5e provided $I^{\alpha\beta}$ is 'inside' Q_4.

In the limit as $I^{\alpha\beta}$ moves onto Q_4, this defines the metric of Minkowski space. Thus de Sitter space is distinguished twistorially from Minkowski space by having a non-simple infinity twistor $I^{\alpha\beta}$.

Chapter 8

Solving the Zero Rest Mass Equations I

In chapter 6 we encountered the g.s.f. condition and showed how it was solved by holomorphic, homogeneous twistor functions. We shall now see that the zero rest mass free field equations can likewise be solved by twistor functions, though the solution is more elaborate. The solution raises a number of questions of interpretation which will be answerable with the aid of sheaf cohomology in the following chapter.

We proceed by means of an example: consider the twistor function

$$f(Z^\alpha) = \frac{1}{(A_\alpha Z^\alpha)(B_\beta Z^\beta)} \tag{8.1}$$

where $A_\alpha = (A_A, A^{A'}), B_\alpha = (B_A, B^{A'})$ are two constant dual twistors.

We aim to calculate a z.r.m. field at the point $x^{AA'}$ in **CM**. Restrict Z^α to the line L_x so that

$$
\begin{aligned}
A_\alpha Z^\alpha &= (iA_A x^{AA'} + A^{A'})\pi_{A'} \equiv \alpha^{A'}\pi_{A'} \\
B_\alpha Z^\alpha &= (iB_A x^{AA'} + B^{A'})\pi_{A'} \equiv \beta^{A'}\pi_{A'}
\end{aligned} \tag{8.2}
$$

and consider the contour integral

$$\varphi(x) = \frac{1}{2\pi i}\oint \frac{1}{(\alpha^{A'}\pi_{A'})(\beta^{B'}\pi_{B'})}\pi_{C'}d\pi^{C'}. \tag{8.3}$$

This is well defined on $\mathbf{P}^1$ since the integrand has total homogeneity zero. Further, there will exist a contour around which to do the integral provided

65

the two poles are distinct, i.e. provided

$$\alpha^{A'}\beta_{A'} \neq 0. \tag{8.4}$$

We may coordinatise $\mathbf{P}^1$ in this case by

$$\pi_{A'} = \alpha_{A'} + z\beta_{A'}$$

when (8.3) becomes

$$
\begin{aligned}
\varphi(x) &= \frac{1}{2\pi i} \oint \frac{dz}{(\alpha^{A'}\beta_{A'})z} \\
&= \frac{1}{\alpha^{A'}\beta_{A'}}.
\end{aligned}
$$

From (8.2), if we suppose that the two dual twistors A_α, B_α meet in the line L_y, we find

$$\alpha^{A'}\beta_{A'} = \frac{1}{2}A_A B^A (x-y)^2$$

where $(x-y)^2 = (x^a - y^a)(x_a - y_a)$. Thus

$$\varphi(x) = \frac{2}{A_A B^A (x-y)^2}. \tag{8.5}$$

It is easy to see, by a direct calculation, that this is a solution of the wave equation. However, it is actually easier to see that, from the definition, it *must* be a solution of the wave equation. If we use the symbol ρ_x in front of a twistor expression to mean that the subsequent twistors are restricted to the line L_x, then the integral (8.3) can be written

$$
\begin{aligned}
\varphi(x) &= \frac{1}{2\pi i} \oint \rho_x f(Z^\alpha)\pi_{C'}d\pi^{C'} \\
&= \frac{1}{2\pi i} \oint f(ix^{AA'}\pi_{A'}, \pi_{A'})\pi_{C'}d\pi^{C'}.
\end{aligned} \tag{8.6}
$$

Now

$$\frac{\partial}{\partial x^{AA'}}\rho_x f = i\pi_{A'}\rho_x \frac{\partial f}{\partial \omega^A}. \tag{8.7}$$

So

$$
\begin{aligned}
\frac{\partial\varphi}{\partial x^{AA'}} &= \frac{1}{2\pi i} \oint \rho_x \frac{\partial f}{\partial \omega^A} i\pi_{A'}\pi_{C'}d\pi^{C'} \\
\frac{\partial^2\varphi}{\partial x^{AA'}\partial x^{BB'}} &= \frac{1}{2\pi i} \oint (-1)\rho_x \frac{\partial^2 f}{\partial \omega^A \partial \omega^B} \pi_{A'}\pi_{B'}\pi_{C'}d\pi^{C'}
\end{aligned}
$$

whence it is clear that

$$\Box \varphi = 0.$$

For the integrand in (8.6) to be well defined on $\mathbf{P}^1$, the function f must be homogeneous of degree -2, or equivalently, a section of the bundle $\mathcal{O}(-2)$ discussed in exercise 3g.

Having solved the wave equation we may consider the other z.r.m. equations. These are solved in a similar way by contour integral formulae (Penrose 1968a, 1969; Penrose and MacCallum 1973):

$$\varphi \underbrace{_{A'\ldots B'}}_{s}(x) = \frac{1}{2\pi i} \oint \pi_{A'} \ldots \pi_{B'} \rho_x f(Z^\alpha) \pi_{C'} d\pi^{C'} \tag{8.8}$$

$$\psi \underbrace{_{A\ldots B}}_{s}(x) = \frac{1}{2\pi i} \oint \rho_x \frac{\partial}{\partial \omega^A} \ldots \frac{\partial}{\partial \omega^B} f(Z^\alpha) \pi_{C'} d\pi^{C'} \tag{8.9}$$

where in the first f is homogeneous of degree $-s - 2$ and in the second f is homogeneous of degree $s - 2$.

To see that these work is a simple application of (8.7). For the first

$$\frac{\partial}{\partial x^{CC'}} \varphi_{A'\ldots B'}(x) = \frac{1}{2\pi i} \oint \pi_{A'} \ldots \pi_{B'} \pi_{C'} \rho_x \frac{\partial f}{\partial \omega^C} \pi_{E'} d\pi^{E'}$$

which is evidently symmetric on $A' \ldots C'$ so that

$$\nabla_C{}^{A'} \varphi_{A'\ldots B'} = 0.$$

The other is similar.

Various properties of this general construction may be seen in the first example. First of all, the field $\varphi(x)$ is singular at points x^a in $\mathbf{CM}$ which are null separated from the point y^a defined by the dual twistors A_α and B_α. If we choose L_y to lie entirely in $\mathbf{PT}^-$ (figure 8.1), then $\varphi(x)$ is nonsingular for all L_x which lie entirely in $\mathbf{PT}^+$, i.e. by exercise 7c, for all x in $\mathbf{CM}^+$, the future tube. A z.r.m. field with the property of being nonsingular on $\mathbf{CM}^+$ is a *positive-frequency* field; thus the notion of positive frequency can be made geometrical. What is needed is that the singularities of $f(Z^\alpha)$ should form two distinct regions on each L_x in $\mathbf{PT}^+$ so that a contour always exists for the integral (8.8). (Recall a positive-frequency field $\psi_{A'\ldots B'}(x)$ with $2n$ indices is said to have *helicity* n while a positive-frequency field $\varphi_{A\ldots B}(x)$ with $2n$ indices has *helicity* $-n$.)

Next we remark that there is a great deal of freedom in the function f in (8.6) subject to it giving the same φ. In particular, we may change f by adding a function h which has singularities on one side of the contour Γ

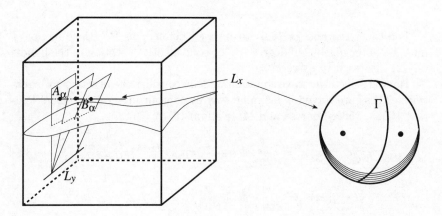

Figure 8.1. For y in the past tube there will be a contour Γ for each x in the future tube making $\varphi(x)$ non-singular.

but is holomorphic on the other, since the contour integral will ignore such h. With f as in (8.1) for example we could take

$$f \to \hat{f} = f + \frac{1}{(B_\alpha Z^\alpha)^2}$$

where B_α is as in figure 8.1. In general we have the freedom

$$f \to f + h - \hat{h} \tag{8.10}$$

where $h, \hat{h}$ are holomorphic respectively on opposite sides of Γ.

There is also the freedom in moving the contour. If we combine these freedoms then the precise relation between $\varphi(x)$ and the class of objects which lead to it via (8.6) becomes rather puzzling! To understand what is happening we shall introduce sheaf cohomology.

Exercises 8

a) A z.r.m. field $\varphi_{A'\dots B'}$ is said to be null if it has only one p.n.d., i.e. if $\varphi_{A'\dots B'} = o_{(A'}\dots o_{B')}$ for some $o_{A'}$. Show that a null field arises from the contour integral (8.8) if $f(Z^\alpha)$ has a simple pole.

As an extension of this, what are the algebraic types of the linearised Weyl spinors obtained from $f_r(Z^\alpha) = (A_\alpha Z^\alpha)^{r-5}(B_\beta Z^\beta)^{-r-1}$ in the cases $r = 0, 1, 2$?

b) If the z.r.m. field $\varphi_{A'...B'}$ is obtained from $f(Z^\alpha)$ show that the fields obtained from $(A_\beta Z^\beta) f(Z^\alpha)$ and $B^\beta \frac{\partial}{\partial Z^\beta} f(Z^\alpha)$ respectively are the result of spin-lowering and spin-raising on $\varphi_{A'...B'}$ (with twistors related to A_α and B^α respectively; see exercise 4d).

c) Solving the Klein–Gordon equation: show that the contour integral

$$\varphi_{A'...B'A...C}(x) = \frac{1}{2\pi i} \oint Z_{A'}...Z_{B'} W_A...W_C \rho_x F(Z^\alpha, W_\alpha) Z_{C'} dZ^{C'} W_D dW^D$$

where

$$\rho_x F(Z^\alpha, W_\alpha) = F(ix^{AA'} Z_{A'}, Z_{A'}, W_{A'}, -ix^{A'A} W_{A'})$$

satisfies the Klein–Gordon equation

$$(\Box + m^2)\varphi_{A'...B'A...C} = 0$$

if

$$I^{\alpha\beta} W_\alpha \frac{\partial}{\partial Z^\beta} I_{\gamma\delta} Z^\gamma \frac{\partial}{\partial W_\delta} F = -\frac{1}{2} m^2 F.$$

In this way a cohomological theory of massive fields can be built up.

d) Calculate the field corresponding to the twistor function

$$\frac{R_\alpha Z^\alpha}{P_\beta Z^\beta (Q_\gamma Z^\gamma)^2} + \frac{S_\alpha Z^\alpha}{(P_\beta Z^\beta)^2 Q_\gamma Z^\gamma}.$$

Show that it is singular on the null cone of the point represented by $P_{[\alpha} Q_{\beta]}$. Show also that, as R_α and S_α vary, these fields form a five-parameter family (rather than eight, as one might naively think!).

Space-time fields arising from twistor functions of the form

$$\frac{(A_\alpha Z^\alpha)^p (B_\alpha Z^\alpha)^q}{(C_\alpha Z^\alpha)^r (B_\alpha Z^\alpha)^s}$$

are known as *elementary states* (Penrose and MacCallum 1973; Hughston and Ward 1979). They are normalisable if $C_{[\alpha} D_{\beta]}$ is in $\mathbf{PT}^+$ or $\mathbf{PT}^-$, and are dense in the space of all fields (Eastwood and Pilato 1991). They constitute a useful alternative to plane waves as basis functions.

Chapter 9

Sheaf Cohomology and Free Fields

We saw in the last chapter that in order to make the relationship between twistor functions and zero rest mass fields precise we seem to need to study certain collections of these (holomorphic) twistor functions. These collections are actually well-known to the mathematician: they arise naturally in the context of sheaf cohomology.

We start by reviewing the essential ideas behind analytic continuation of complex functions. This will quickly lead us to discuss Riemann surfaces and sheaves.

Consider an analytic function $f : D \to \mathbf{C}$, where $D \subset \mathbf{C}$ is a domain (that is, an open connected set). Under what circumstances can we *extend* f so that it is defined (and still analytic) on some larger set than D? Well, as usual, the analyticity allows us to push out the boundaries of D until the 'natural boundary' of f (a singularity of some sort) is reached. It frequently happens that in doing this the function f gets continued *back* to a point $z \in D$, but takes a different value there from its original one. The archetypical example, of course, is the logarithm. We therefore seem to be obliged to study multivalued functions, unless we can somehow absorb this behaviour into the *space* on which f is defined.

More precisely, we define a *function element* to be a pair (f, D) where f is analytic on the domain D. We say that two function elements (f_1, D_1) and (f_n, D_n) are *equivalent* (written $(f_1, D_1) \sim (f_n, D_n)$) if there is a sequence of function elements

$$(f_2, D_2), (f_3, D_3), \ldots, (f_{n-1}, D_{n-1})$$

71

such that $D_i \cap D_{i+1} \neq \emptyset$ and $f_i = f_{i+1}$ on $D_i \cap D_{i+1}$ for $i = 1, \ldots, n-1$.
This equivalence relation encapsulates the notion of analytic continuation:
the equivalence *classes* are the *complete analytic functions*. It is these c.a.
functions which are liable to be multivalued, so our aim is given a c.a.f. to
construct a space R so that the given c.a.f. can be regarded as an ordinary
function from R to $\mathbf{C}$. This space R will be the *Riemann surface* of the
c.a.f.; it will be a one-dimensional complex manifold. Broadly speaking,
R is constructed as follows. Suppose the c.a.f. is the equivalence class
consisting of the function elements (f_α, D_α) for α in some labelling set A.
If we took the union of all the D_α, we would obtain the domain in $\mathbf{C}$ on
which our c.a.f. is liable to be multivalued. So we take instead the *disjoint
union* of all the D_α. That is, we regard D_α and D_β as disjoint even if they
have some or all of their points in common. So if D_α and D_β *do* have a
point in common, there are now two copies of it, say z_α and z_β. Clearly this
new space is much too big. We squash it down to the right size and obtain
our space R by identifying z_α with z_β wherever there is a neighbourhood
of z on which $f_\alpha = f_\beta$. Actually, for many c.a.f.s (like $z^{\frac{1}{2}}$ for example) we
could simply have identified z_α with z_β wherever $f_\alpha(z) = f_\beta(z)$. Then R
is the same as the *graph* of the function $f : R = \{(z, w) \in \mathbf{C}^2 : z = w^2\}$;
see figure 9.1. In these cases it is easy to see that R is a one-dimensional
complex manifold. Unfortunately, however, this construction would not
work for the function $w = (z - i)\sqrt{z}$. The graph of this function only has
one point $(i, 0)$ above $z = i$. Its Riemann surface however, has two, one for
each branch of the function. This is why we identify z_α with z_β if there is
a *neighbourhood* of z on which $f_\alpha = f_\beta$, in our construction of R.

So far we have just been considering one complete analytic function and
its corresponding Riemann surface. Our aim in the first part of this chapter
is to find a way of describing the space of *all* c.a.f.s. In some sense therefore
we want to study the union of all Riemann surfaces. In order to make this
precise we need the notion of a germ.

Given a point z and a function element (f, D) such that $z \in D$, the
germ $[f, z]$ of f at z is the set of all function elements (f_i, D_i) such that
$z \in D_i$ and there is a neighbourhood of z on which $f_i = f$. Given a germ

$$[f, z] = \{(f_i, D_i) : i \in I\}$$

we can recover the point $z (= \cap D)$, the value $f(z)$ and the c.a.f. from which
the germ is taken.

The set of all germs, for all $z \in \mathbf{C}$, is denoted $\mathcal{O}$ and called the *sheaf* of
germs of holomorphic functions on $\mathbf{C}$. We see next that $\mathcal{O}$ has a natural
(Hausdorff) topology, and in fact is a one-dimensional complex manifold.

To specify a topology on $\mathcal{O}$, we have to say which subsets of $\mathcal{O}$ are open.

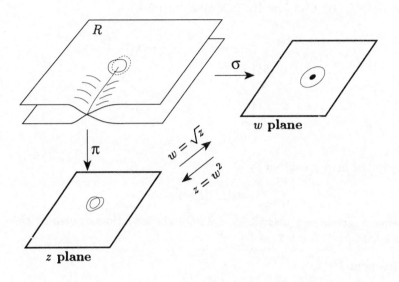

Figure 9.1.

Given a function element (f, D) there is a corresponding subset of $\mathcal{O}$:

$$[f, D] = \{[f, z] : z \in D\}.$$

We choose these to be the basic open subsets of $\mathcal{O}$. (It is easily checked that they satisfy the axioms for a Hausdorff topology.) An arbitrary open set in $\mathcal{O}$ is then simply the *union* of a collection of the basic open sets. Now the map $\pi : \mathcal{O} \to \mathbf{C}$ given by $\pi([f, z]) = z$ has some very nice properties. It maps the basic open sets $[f, D]$ of $\mathcal{O}$ *homeomorphically* onto the open sets D of $\mathbf{C}$. (Again this is easily checked: see exercise 9a.)

   In particular, therefore, we can use $\pi|_{[f, D]}$ as a *chart* and in so doing we can give the connected components of $\mathcal{O}$ the structure of a one-dimensional complex manifold. (It is not a connected manifold though.) These connected components are the various Riemann surfaces of the c.a.f.s on $\mathbf{C}$. We see this by considering two function elements (f, U) and (g, V). If $z \in U$ and $w \in V$ then it can be shown that a path γ in $\mathcal{O}$ from the germ $[f, z]$ to the germ $[g, w]$ corresponds to analytic continuation from the function element (f, U) to (g, V). So γ exists iff (f, U) and (g, V) are equivalent.

   This is our description of the sheaf $\mathcal{O}$, except for one last point. We have lumped all the function elements together into c.a.f.s, and all the c.a.f.s together into one (large) space $\mathcal{O}$. Given $\mathcal{O}$, how do we recover a function element (f, D)? Consider the following diagram:

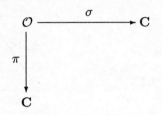

where the map σ is given by

$$\sigma([f, z]) = f(z).$$

Define a *section* of π over the domain D to be a continuous map $s : D \to \mathcal{O}$ such that $\pi(s(z)) = z \, \forall z \in D$.

Theorem 1

For any domain D there is a 1-1 correspondence between function elements (f, D) and sections of π over D, such that $f(z) = \sigma(s(z)) \, \forall z \in D$.

Proof

Given a function element (f, D) define

$$s(z) = [f, z] \, \forall z \in D.$$

This is the required section, and clearly

$$f(z) = \sigma(s(z)).$$

Given a section s of π over D, and a point z in D,

$$s(z) = [g, z]$$

where g is holomorphic in a domain U containing z. We define $f(z) = \sigma(s(z))$, and we have to show that f is holomorphic. The continuity of s implies that the open set $[g, U]$ in $\mathcal{O}$ is mapped (homeomorphically) to U by $\pi \, (= s^{-1})$. So $\forall z \in U \cap D$:

$$f(z) = \sigma(s(z)) = \sigma([g, z]) = g(z)$$

and hence f is holomorphic on $U \cap D$. We repeat this argument for different sets U until D is covered, and deduce that $f = \sigma \circ s$ is holomorphic on D.

□

From the point of view of one-variable complex analysis $\mathcal{O}$ is in itself a desirable goal. For our purposes however it is actually just a very useful example. After all, we need to study holomorphic functions on $\mathbf{PT}^+$, not on $\mathbf{C}$. So we shall summarise our discussion of $\mathcal{O}$ by writing the *axioms* for a sheaf, in preparation for looking at some more exotic examples. (The first of these axioms will be no surprise, but we have not made explicit the motivation for the others. In fact it is easy to see that the *stalk* $\mathcal{O}_z = \pi^{-1}(z)$ of $\mathcal{O}$ over z is a commutative ring.)

Let X be a topological space. A *sheaf* S over X is a topological space together with a mapping $\pi : S \to X$ which satisfies:

i) π is a local homeomorphism;

ii) the stalks $S_x = \pi^{-1}(x)$ are abelian groups;

iii) the group operations are continuous.

The sections of S over an open set U in X form an abelian group $S(U)$. The elements of this group are the function elements we were studying in the context of the sheaf $\mathcal{O}$, but in general they need not be analytic functions. Indeed if X is a C^∞ differentiable manifold we could choose our function elements to be C^∞ differentiable functions on open domains in X, or p-forms, or closed p-forms. We obtain in this way the sheaves $\mathcal{A}^0, \mathcal{A}^p$, and $\mathcal{Z}^p$ on X. In twistor theory, though, the spaces X on which we want to construct sheaves are complex manifolds. Then our function elements could for example be holomorphic functions, holomorphic p-forms, C^∞ forms of type p, q, or $\bar{\partial}$ closed C^∞ forms of type p, q, and we would obtain the sheaves $\mathcal{O}, \Omega^p, \mathcal{A}^{p,q}$, and $\mathcal{Z}^{p,q}$. (Forms of type p, q and the $\bar{\partial}$ operator are introduced in exercise 9d and are used in chapter 14.)

We can relate all these sheaves to each other using sheaf homomorphisms. Suppose S and T are sheaves over the same space X. The continuous map

$$\varphi : S \to T$$

is a *sheaf homomorphism* if it preserves stalks

$$\varphi : S_x \to T_x \, \forall x \in X$$

and is a group homomorphism on each stalk. We shall see an example in a moment. First we define the usual concepts of kernel and image:

$$\ker \varphi = \{s \in S : \varphi(s) = 0 \text{ (the zero element in } T_x, \text{ where } s \in S_x)\}$$
$$\operatorname{im} \varphi = \{t \in T : t = \varphi(s) \text{ for some } s \in S\}.$$

A *short exact sequence* of sheaves over a space X is a sequence of sheaves and sheaf homomorphisms

$$0 \to \mathcal{A} \overset{\varphi}{\to} \mathcal{B} \overset{\psi}{\to} \mathcal{C} \to 0$$

(where 0 is the zero sheaf) which is exact at each stage:

$$\begin{aligned}
\ker \varphi &= 0 \\
\operatorname{im} \varphi &= \ker \psi \\
\operatorname{im} \psi &= \mathcal{C}.
\end{aligned}$$

An alternative way of looking at this is to observe that the sequence above is exact if $\forall x \in X$ the sequence of stalks

$$0_x \to \mathcal{A}_x \to \mathcal{B}_x \to \mathcal{C}_x \to 0_x$$

is exact (as a sequence of abelian groups and group homomorphisms). This emphasises that exactness is a *local* property, as we can also see clearly in the following example.

Let X be a complex manifold, $\mathbf{Z}$ be the constant sheaf of integers over X (the space of c.a.f.s from X to $\mathbf{Z}$), and $\mathcal{O}^*$ be the sheaf of germs of non-vanishing holomorphic functions on X in which the abelian group operation is *multiplication*. Then

$$0 \to \mathbf{Z} \overset{i}{\to} \mathcal{O} \overset{e}{\to} \mathcal{O}^* \to 0$$

is a short exact sequence, where i is injection and

$$e([f, z]) = [\exp(2\pi i f), z].$$

The sequence is obviously exact at $\mathbf{Z}$ and $\mathcal{O}$, but exactness at $\mathcal{O}^*$ is more interesting. Given a germ $[g, z]$ we need to find another germ $[f, z]$ such that

$$e([f, z]) = [g, z].$$

For any $[g, z]$ we can find a simply connected neighbourhood N of z and define a branch of $\log g$ on N (recall that $g \neq 0$). Then we choose $f = \frac{1}{2\pi i} \log g$.

In contrast, let us now take an open set U in X and consider the sequence of abelian groups

$$0 \to \mathbf{Z}(U) \to \mathcal{O}(U) \to \mathcal{O}^*(U) \to 0$$

(where $\mathcal{S}(U)$ is the group of sections of the sheaf $\mathcal{S}$ over the set U, remember). This sequence is also exact at $\mathbf{Z}(U)$ and $\mathcal{O}(U)$, but *not* at $\mathcal{O}^*(U)$, precisely because the required logarithm may not be available.

This is only one (perhaps the best known) of many examples illustrating the ability of sheaf theory to express in an elegant way the distinction between local and non-local information. We can go further however. We shall see presently that sheaf *cohomology* theory enables us to put a *measure* on that distinction: we shall be able to decide under what circumstances, and by *how much*, the sequence above fails to be exact at $\mathcal{O}^*(U)$.

Čech Cohomology

Consider a sheaf $\mathcal{S}$ over a topological space X, and an open cover $\{U_i\}$ of X. (That is, the U_i are open sets in X such that $\cup_i U_i = X$.) We will need to simplify our notation a little by abbreviating $U_i \cap U_j$ to U_{ij}. Our first definition is that of a *p-cochain*. This is a collection of sections $f_{i_0 \ldots i_p} \in \mathcal{S}(U_{i_0 \ldots i_p})$, one for each non-empty $p+1$-fold intersection of the sets U_i, and completely skew-symmetric:

$$f_{i_0 \ldots i_p} = f_{[i_0 \ldots i_p]}.$$

We should refer to this p-cochain as $\{f_{i_0 \ldots i_p}\}$ of course, but we often simply say 'the p-cochain $f_{i_0 \ldots i_p}$'.

This definition is slightly less intimidating when one learns that in practice p is always between 0 and 3! For example, a 0-cochain is a collection of sections $f_i \in \mathcal{S}(U_i)$ and a 1-cochain is a collection of sections $f_{ij} \in \mathcal{S}(U_{ij})$ such that $f_{ij} = -f_{ji}$.

The set of all p-cochains has an abelian group structure (inherited from the sheaf) and is denoted

$$C^p(\{U_i\}; \mathcal{S}).$$

Our next task is to define the *coboundary map*

$$\delta_p : C^p \to C^{p+1}$$

and we do this first for $p = 0$. Given a 0-cochain f_i we define a 1-cochain f_{ij} as follows:

$$f_{ij} = \rho_j f_i - \rho_i f_j,$$

where $\rho_i f_j$ is the restriction of f_j to the set U_{ij}. Let us streamline our notation again and write this as

$$f_{ij} = 2\rho_{[j} f_{i]}.$$

Then

$$\delta_0(\{f_i\}) = \{2\rho_{[j} f_{i]}\}$$

is the coboundary map on C^0:

$$\delta_0 : C^0(\{U_i\}; \mathcal{S}) \to C^1(\{U_i\}; \mathcal{S}).$$

We can see straight away that

$$\ker \delta_0 = \mathcal{S}(X).$$

The coboundary map $\delta_p : C^p \to C^{p+1}$ is defined analogously:

$$\delta_p(\{f_{i_0 \ldots i_p}\}) = \{(p+1)\rho_{[i_{p+1}}f_{i_0 \ldots i_p]}\}. \tag{9.1}$$

There are special names for the kernels and images of these coboundary maps:

$$Z^p(\{U_i\}; \mathcal{S}) = \ker \delta_p$$
$$B^p(\{U_i\}; \mathcal{S}) = \operatorname{im} \delta_{p-1}$$

and the elements of Z^p are called *p-cocycles*, while the elements of B^p are called *p-coboundaries*. Again, Z^p and B^p are abelian groups, and it is easy to see from (9.1) that

$$\delta_{p+1} \circ \delta_p = 0 \tag{9.2}$$

so that B^p is a (normal) subgroup of Z^p.

The pth Čech cohomology group of the cover $\{U_i\}$ with coefficients in the sheaf $\mathcal{S}$ is then defined to be

$$\check{H}^p(\{U_i\}; \mathcal{S}) = \frac{Z^p(\{U_i\}; \mathcal{S})}{B^p(\{U_i\}; \mathcal{S})}. \tag{9.3}$$

The best way to become familiar with these definitions is to study an example, of course. We look at a very simple and useful one, namely $\mathbf{CP}^1$ covered by the two sets

$$U_0 = \{[\pi_{0'}, \pi_{1'}] \in \mathbf{CP}^1 : \pi_{0'} \neq 0\}$$
$$U_1 = \{[\pi_{0'}, \pi_{1'}] \in \mathbf{CP}^1 : \pi_{1'} \neq 0\} \tag{9.4}$$

and we choose the sheaf $\mathcal{S}$ to be $\mathcal{O}$. Then

$$\check{H}^0(\{U_i\}; \mathcal{O})$$

is the space of pairs of functions f_i analytic on U_i $(i = 0, 1)$ satisfying

$$\rho_1 f_0 = \rho_0 f_1.$$

In other words f_0 and f_1 agree on the overlap U_{01} and therefore define a complete analytic function on U_0 and U_1, which is a global analytic function on $\mathbf{CP}^1$. By Liouville's Theorem this must be constant, so

$$\check{H}^0(\{U_i\}; \mathcal{O}) = \mathbf{C}. \tag{9.5}$$

Next we consider

$$\check{H}^1(\{U_i\}; \mathcal{O}).$$

This is the space of functions f_{01} in U_{01} modulo coboundaries. (There is no cocycle condition because with this cover there is no triple intersection.) The domain U_{01} is an annulus, so these functions f_{01} are precisely those having a Laurent expansion, which we write in the form

$$f_{01}(\zeta) = \sum_{n=1}^{\infty} b_n \zeta^{-n} - \sum_{n=0}^{\infty} a_n \zeta^n$$

(where $\zeta = \frac{\pi_{0'}}{\pi_{1'}}$). Let

$$g_0(\zeta) = \sum_{n=1}^{\infty} b_n \zeta^{-n}$$

and

$$g_1(\zeta) = \sum_{n=0}^{\infty} a_n \zeta^n.$$

Then g_i converges on U_i $(i = 0, 1)$ and

$$f_{01} = \rho_1 g_0 - \rho_0 g_1$$

so that f_{01} is a coboundary. Hence

$$\check{H}^1(\{U_i\}; \mathcal{O}) = 0. \tag{9.6}$$

These results are not quite what we will need though. Local sections of the sheaf $\mathcal{O}$ are analytic functions homogeneous of degree *zero* in the homogeneous coordinates $[\pi_{0'}, \pi_{1'}]$ for $\mathbf{CP}^1$. How do we describe analytic functions with *non*-zero homogeneity? They cannot really be *functions* on $\mathbf{CP}^1$ of course, and indeed we saw in exercise 3g that they are sections of certain line bundles $\mathcal{O}(n)$ on $\mathbf{CP}^1$.

With this in mind we calculate

$$\check{H}^0(\{U_i\}; \mathcal{O}(-1)).$$

This is quite easy, because the elements here are global sections and in this case there cannot be any because a function with negative homogeneity must have a singularity somewhere. So

$$\check{H}^0(\{U_i\}; \mathcal{O}(-1)) = 0. \tag{9.7}$$

Indeed, the same argument tells us that

$$\check{H}^0(\{U_i\}; \mathcal{O}(n)) = 0 , \; n < 0.$$

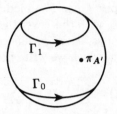

Figure 9.2. The splitting formula.

A global function homogeneous of degree zero represents a function on the complex plane with no poles and finite at infinity. By Liouville's Theorem such a function is necessarily constant so

$$\check{H}^0(\{U_i\}; \mathcal{O}(0)) = \mathbf{C}.$$

Finally a global function $g(\pi_{A'})$ homogeneous of degree $n > 0$ has n-th derivative $\frac{\partial^n g}{\partial \pi_{A'} \dots \partial \pi_{B'}}$ which is homogeneous of degree zero and hence constant. Thus

$$g = \varphi^{A' \dots B'} \underbrace{\pi_{A'} \dots \pi_{B'}}_{n}$$

i.e. g is a polynomial and

$$\check{H}^0(\{U_i\}; \mathcal{O}(n)) = \mathbf{C}^{n+1} \ , \ n > 0.$$

Another group we will need is

$$\check{H}^1(\{U_i\}; \mathcal{O}(-1)).$$

This is the space of functions f_{01} analytic on U_{01} and homogeneous of degree -1 in the coordinates $[\pi_{0'}, \pi_{1'}]$, modulo coboundaries. We define (following Sparling : see § 2.4 of Hughston and Ward 1979)

$$h_i(\pi_{0'}, \pi_{1'}) = \frac{1}{2\pi i} \int_{\Gamma_i} \frac{f_{01}(\lambda, 1) d\lambda}{\lambda \pi_{1'} - \pi_{0'}} \ , \ i = 0, 1 \tag{9.8}$$

where $\lambda = \frac{\lambda^0}{\lambda^1}$; the integrand is homogeneous of degree zero in λ^0, λ^1 as required. The contours Γ_0 and Γ_1 are drawn in figure 9.2 so that $\Gamma_0 - \Gamma_1$

is equivalent to a contour surrounding the point

$$(\lambda^0, \lambda^1) = (\pi_{0'}, \pi_{1'})$$

and then by the Cauchy Integral Formula

$$
\begin{aligned}
h_0(\pi_{0'}, \pi_{1'}) - h_1(\pi_{0'}, \pi_{1'}) &= \frac{1}{2\pi i} \int_{\Gamma_0 - \Gamma_1} \frac{f_{01}(\lambda, 1) d\lambda}{\lambda \pi_{1'} - \pi_{0'}} \\
&= f_{01}(\pi_{0'}, \pi_{1'}).
\end{aligned}
$$

This exhibits f_{01} as a coboundary and so

$$\check{H}^1(\{U_i\}; \mathcal{O}(-1)) = 0. \tag{9.9}$$

Of course we could also have shown this using the Laurent series as before, but that would have hidden the fact, which will be of some significance to us, that an integral is involved in representing f_{01} as a coboundary. (Recall that the Laurent series coefficients are determined by the Cauchy Integral Formula.)

Now for $\check{H}^1(\{U_i\}; \mathcal{O}(n))$ with $n \geq 0$. Given $f_{01}(\pi_{A'})$ with non-negative homogeneity, we may divide it by a polynomial of degree $(n+1)$ to give $g_{01}(\pi_{A'})$ of homogeneity -1. This can be split as above. Then multiplying by the polynomial we reduce f_{01} to a coboundary. Thus

$$\check{H}^1(\{U_i\}; \mathcal{O}(n)) = 0, \ n \geq -1.$$

For $\check{H}^1(\{U_i\}; \mathcal{O}(-n))$ with $n > 1$ we have

$$
\begin{aligned}
f_{01}(\pi_{0'}, \pi_{1'}) &= \frac{1}{(\pi_{1'})^n} f_{01}(\zeta, 1) \\
&= \frac{1}{(\pi_{1'})^n} \sum_{-\infty}^{\infty} a_r \zeta^r \\
&= \frac{1}{(\pi_{1'})^n} \left(\sum_{-\infty}^{-n} a_r \zeta^r + \sum_{-n+1}^{-1} a_r \zeta^r + \sum_{0}^{\infty} a_r \zeta^r \right) \\
&= \sum_{n}^{\infty} a_{-r} \frac{(\pi_{1'})^{r-n}}{(\pi_{0'})^r} + \sum_{0}^{\infty} a_r \frac{(\pi_{0'})^r}{(\pi_{1'})^{n+r}} + g_{01}(\pi_{A'})
\end{aligned}
$$

where

$$g_{01}(\pi_{A'}) = \sum_{1}^{n-1} \frac{a_{-r}}{(\pi_{0'})^r (\pi_{1'})^{n-r}}. \tag{9.10}$$

n	$\cdots -4$	-3	-2	-1	0	1	$2 \cdots$
$\check{H}^0(\{U_i\}; \mathcal{O}(n))$	0	0	0	0	$\mathbf{C}^1$	$\mathbf{C}^2$	$\mathbf{C}^3$
$\check{H}^1(\{U_i\}; \mathcal{O}(n))$	$\mathbf{C}^3$	$\mathbf{C}^2$	$\mathbf{C}^1$	0	0	0	0

Table 9.1.

So f_{01} is cohomologous to g_{01} (i.e. it differs from g_{01} by a coboundary) and g_{01} cannot be simplified further. There are $n - 1$ complex parameters in g_{01} so

$$\check{H}^1(\{U_i\}; \mathcal{O}(-n)) \cong \mathbf{C}^{n-1}, \ n \geq 2.$$

We may set these results out as in table 9.1.

Evidently there is some sort of duality here. In fact, as we shall see,

$$\check{H}^1(\{U_i\}; \mathcal{O}(-n-2)) = \check{H}^0(\{U_i\}; \mathcal{O}(n))^* \tag{9.11}$$

where the star denotes vector space dual.

This a special instance of *Serre duality* (Wells 1980; Field 1982). Here it is expressed by another contour integral formula: if

$$g(\pi_{A'}) = \varphi^{A' \ldots B'} \underbrace{\pi_{A'} \ldots \pi_{B'}}_{n} \in \check{H}^0(\{U_i\}; \mathcal{O}(n))$$

and $f(\pi_{A'}) \in \check{H}^1(\{U_i\}; \mathcal{O}(-n-2))$ is given by a representative like (9.10) then consider

$$(f, g) = \frac{1}{2\pi i} \oint g(\pi_{A'}) f(\pi_{A'}) \pi_{C'} d\pi^{C'} \tag{9.12}$$

where the integral is round the equator. Evidently this expression is complex linear and non-degenerate and depends only on the cohomology class of f. Thus it defines the duality (9.11).

A suggestive way of writing (9.12) is as

$$(f, g) = \psi_{A' \ldots B'} \varphi^{A' \ldots B'}$$

where

$$\psi_{A' \ldots B'} = \frac{1}{2\pi i} \oint \pi_{A'} \ldots \pi_{B'} f(\pi_{D'}) \pi_{C'} d\pi^{C'} \tag{9.13}$$

when it is clear that f determines $\psi_{A' \ldots B'}$, an element of the dual of $\check{H}^0(\{U_i\}; \mathcal{O}(n))$. Now the contour integral formula (8.8) can be recognized as an instance of Serre duality.

In (8.8) of course, f was a twistor function. We can now see how to interpret it cohomologically. We cover $\mathbf{PT}^+$ by the open sets

$$U_1 = \{Z^\alpha : Z^\alpha A_\alpha \neq 0\}$$
$$U_2 = \{Z^\alpha : Z^\alpha B_\alpha \neq 0\}$$

and then if f is the particular example (8.1):

$$f(Z^\alpha) = f_{12}(Z^\alpha) = \frac{1}{(Z^\alpha A_\alpha)(Z^\alpha B_\alpha)}$$

we have a 1 cochain. In fact, f_{12} is a cocycle since there are no triple intersections in the cover $\{U_1, U_2\}$. Further, the freedom (8.10) which we found in f precisely corresponds to changing f by a coboundary. Thus the twistor function f is properly thought of as an element of $\check{H}^1(\{U_i\}; \mathcal{O}(-2))$.

Since the resulting space-time field φ was positive-frequency we are led to conjecture that

$$\check{H}^1(\mathbf{PT}^+; \mathcal{O}(-2)) \cong \{\text{positive frequency-solutions of the wave equation}\}$$

or more generally that

$$\check{H}^1(\mathbf{PT}^+; \mathcal{O}(n)) \cong \{\text{positive-frequency z.r.m. fields of helicity } -n-2\}.$$
$$(9.14)$$

We shall see in chapter 10 that this is indeed the case.

An unsatisfactory feature of all these calculations has been the apparent dependence on the cover $\{U_i\}$ of $\mathbf{CP}^1$. It would be very reasonable, for example, to wonder how to compare two cocycles f_{ij} and g_{ab} defined with respect to the different covers $\{U_i\}$ and $\{V_a\}$ of the same space X. (We imagine f_{ij} and g_{ab} to be local sections of the same *sheaf* $\mathcal{S}$ of course.)

To do this, we need the notion of a *refinement*. The cover $\{W_\alpha\}$ is a refinement of $\{U_i\}$ if each of the sets W_α is a subset of (at least) one of the sets U_i. In other words, if the indexing sets for $\{U_i\}$ and $\{W_\alpha\}$ are I and A then there is a map

$$r : A \to I$$

such that

$$W_\alpha \subset U_{r(\alpha)} \text{ for all } \alpha \in A.$$

This induces a map in the other direction on cochains:

$$r^* : C^p(\{U_i\}; \mathcal{S}) \to C^p(\{W_\alpha\}; \mathcal{S}).$$

In the case $p = 1$ for example,

$$(r^* f)_{\alpha\beta} = f_{r(\alpha)r(\beta)}|W_{\alpha\beta}.$$

In fact a little work (not deep, just fiddly) shows that r^* commutes with δ, so that it is actually a map on cohomology classes:

$$r^* : \check{H}^p(\{U_i\}; \mathcal{S}) \to \check{H}^p(\{W_\alpha\}; \mathcal{S}).$$

Now suppose that $\{W_\alpha\}$ is simultaneously a refinement of $\{U_i\}$ and $\{V_a\}$ (with refining maps r^* and ρ^* say). Then to compare our cocycles f and g we first take their cohomology *classes* in $\check{H}^1(\{U_i\}; \mathcal{S})$ and $\check{H}^1(\{V_a\}; \mathcal{S})$, denoted by $[f]$ and $[g]$, and then compare $r^*[f]$ and $\rho^*[g]$ in $\check{H}^1(\{W_\alpha\}; \mathcal{S})$. If they are equal we regard $[f]$ and $[g]$ as being *equivalent*. This idea of equivalence applies to the disjoint union (over *all* covers $\{U_i\}$ of X) of $\check{H}^1(\{U_i\}; \mathcal{S})$. In other words, given *any* two elements of $\check{H}^1$, with respect to *any* two covers, we can ask whether there exists a common refinement of these covers on which the given cohomology classes agree.

Finally, we can define

$$\check{H}^1(X; \mathcal{S})$$

as the set of equivalence *classes* of this new equivalence relation. ($\check{H}^p(X; \mathcal{S})$ is defined similarly of course.)

There is no denying that this is messy though, and in practice one always uses a particular cover. Fortunately for the adherents of the Čech approach to cohomology (who would otherwise be fewer in number) there are theorems guaranteeing that under certain conditions on the cover $\{U_i\}$ and the sheaf $\mathcal{S}$

$$\check{H}^p(\{U_i\}; \mathcal{S}) = \check{H}^p(X; \mathcal{S}).$$

These covers are called *Leray* covers (see Field 1982), and our cover $\{U_0, U_1\}$ of $\mathbf{CP}^1$ is one such, so that we have actually showed (in equations (9.5)–(9.9)) that

$$\check{H}^0(\mathbf{CP}^1; \mathcal{O}) = \mathbf{C}, \tag{9.15}$$

$$\check{H}^1(\mathbf{CP}^1; \mathcal{O}) = 0, \tag{9.16}$$

$$\check{H}^0(\mathbf{CP}^1; \mathcal{O}(-1)) = 0, \tag{9.17}$$

$$\check{H}^1(\mathbf{CP}^1; \mathcal{O}(-1)) = 0. \tag{9.18}$$

[There is another way of defining sheaf cohomology which avoids any awkwardness over covers. A *resolution* of a sheaf $\mathcal{S}$ is a sequence of sheaves and sheaf homomorphisms

$$0 \to \mathcal{S} \xrightarrow{\theta} \mathcal{A}^0 \xrightarrow{\phi_0} \mathcal{A}^1 \xrightarrow{\phi_1} \mathcal{A}^2 \to \cdots \tag{9.19}$$

which is exact at each stage.

By taking the corresponding groups of sections we get the sequence

$$0 \to \mathcal{S}(X) \xrightarrow{\theta^*} \mathcal{A}^0(X) \xrightarrow{\phi_0^*} \mathcal{A}^1(X) \xrightarrow{\phi_1^*} \cdots \qquad (9.20)$$

This is not necessarily exact, but it is a *complex* in that im $\phi_n^* \subset \ker \phi_{n+1}^*$. We may therefore take its cohomology:

$$H^p(X;\mathcal{S}) = \frac{\ker \phi_p^*}{\operatorname{im} \phi_{p-1}^*}. \qquad (9.21)$$

To relate cohomology in this sense to the earlier notion of Čech cohomology one more definition is needed, that of a fine sheaf. A sheaf $\mathcal{A}$ is *fine* if it admits partitions of unity i.e. if for any locally finite covering $\{U_i\}$ of X there exist homomorphisms $h_i : \mathcal{A} \to \mathcal{A}$ such that

i) $h_i(S_x) = 0$ for $x \notin W_i$, where W_i is a closed subset of U_i

ii) $\sum_i h_i = 1$ (note that this is a finite sum since the cover is locally finite).

Typically, sheaves of C^∞ objects are fine, while sheaves of analytic objects are not.

Now there is a theorem that if the resolution (9.19) is *fine*, i.e. all of the sheaves $\mathcal{A}^i$ are fine, then

$$H^p(X;\mathcal{S}) = \check{H}^p(X;\mathcal{S}). \qquad (9.22)$$

Evidently, this gives an effective way of computing cohomology groups as long as fine resolutions can be found. See exercise 9d and chapter 14.]

We promised before introducing cohomology that it would enable us to measure the extent to which *local* information about functions on the space X (expressed in a short exact sequence of sheaves on X) can be made global. We now turn to this question, our last piece of theory before we study massless free fields in terms of sheaf cohomology on twistor space.

We noted before that although the sheaf sequence

$$0 \to \mathbf{Z} \to \mathcal{O} \to \mathcal{O}^* \to 0 \qquad (9.23)$$

is exact, the sequence of sections

$$0 \to \mathbf{Z}(X) \to \mathcal{O}(X) \to \mathcal{O}^*(X) \to 0$$

or equivalently

$$0 \to \check{H}^0(X;\mathbf{Z}) \to \check{H}^0(X;\mathcal{O}) \to \check{H}^0(X;\mathcal{O}^*) \to 0$$

may not be exact at $\mathcal{O}^*(X)$. If however, the set X were simply-connected, then we would always be able to define logarithms and the sequence *would* be exact at $\mathcal{O}^*(X)$. In fact the condition for exactness is *precisely* that X should be simply-connected, or in other words that the de Rham cohomology group $H^1(X; \mathbf{Z})$ should vanish. How do we establish a result like this?

Theorem 2

Given a short exact sequence of sheaves

$$0 \to \mathcal{A} \xrightarrow{\alpha} \mathcal{B} \xrightarrow{\beta} \mathcal{C} \to 0$$

on a topological space X, there is a long exact sequence in cohomology as follows:

$$0 \to \check{H}^0(X; \mathcal{A}) \xrightarrow{\alpha^*} \check{H}^0(X; \mathcal{B}) \xrightarrow{\beta^*} \check{H}^0(X; \mathcal{C}) \xrightarrow{\delta^*} \check{H}^1(X; \mathcal{A}) \xrightarrow{\alpha^*} \check{H}^1(X; \mathcal{B}) \xrightarrow{\beta^*} \ldots$$

Proof

From the short exact sequence we can construct a diagram

$$
\begin{array}{ccccccccc}
 & & \downarrow & & \downarrow & & \downarrow & & \\
0 \to & & C^p(X; \mathcal{A}) & \xrightarrow{\alpha} & C^p(X; \mathcal{B}) & \xrightarrow{\beta} & C^p(X; \mathcal{C}) & \to 0 \\
 & & \downarrow \delta & & \downarrow \delta & & \downarrow \delta & & \\
0 \to & & C^{p+1}(X; \mathcal{A}) & \xrightarrow{\alpha} & C^{p+1}(X; \mathcal{B}) & \xrightarrow{\beta} & C^{p+1}(X; \mathcal{C}) & \to 0 \\
 & & \downarrow & & \downarrow & & \downarrow & &
\end{array}
$$

in which the rows are exact, the squares commute, and the columns are 'complexes' (i.e. $\delta_{p+1} \circ \delta_p = 0$ as in (9.2)). Because the squares commute the definitions of α^* and β^* are clear and because the rows are exact we have

$$\ker \beta^* = \operatorname{im} \alpha^*.$$

The crucial part of the proof though is the construction of the map δ^*, called the *connecting map*. We start with an element $[f]$ of $\check{H}^p(X; \mathcal{C})$ with representative cocycle f. (Note that we are suppressing the indices.) Then $f \in C^p(X; \mathcal{C})$ and $\delta f = 0$. Because β is onto, $f = \beta g$ for some $g \in C^p(X; \mathcal{B})$, and $\beta \delta g = \delta \beta g = \delta f = 0$. So $\delta g \in C^{p+1}(X; \mathcal{B})$ is in the kernel of β, which is the same as the image of α, and there must be an element $h \in C^{p+1}(X; \mathcal{A})$ with $\alpha h = \delta g$. We define $\delta^*[f] = [h]$, but we should check that it is well defined. Suppose we had chosen $g' \in C^p(X; \mathcal{B})$ different from g but with

$$f = \beta g = \beta g'.$$

Then $\beta(g - g') = 0$ so that $g - g' = \alpha k$ and $h - h' = \delta(\alpha k)$ which means that $[h] = [h']$. Similarly, if we had chosen a different representative f' for $[f]$ we would have obtained the same class $[h]$.

All that remains is to check that

$$\ker \delta^* = \operatorname{im} \beta^* \text{ and } \operatorname{im} \delta^* = \ker \alpha^*$$

which we leave as a pastime for the reader.

□

Our short exact sequence (9.23) yields

$$0 \to \check{H}^0(X; \mathbf{Z}) \to \check{H}^0(X; \mathcal{O}) \to \check{H}^0(X; \mathcal{O}^*) \to \check{H}^1(X; \mathbf{Z}) \to$$
$$\check{H}^1(X; \mathcal{O}) \to \dots$$

so all we have to do is note that $\check{H}^1(X; \mathbf{Z})$ is isomorphic to the de Rham cohomology group $H^1(X; \mathbf{Z})$ (Bott and Tu 1982) to get the result claimed earlier.

In addition though, there is some more interesting information further along this same long exact sequence:

$$\dots \to \check{H}^1(X; \mathcal{O}) \xrightarrow{e^*} \check{H}^1(X; \mathcal{O}^*) \xrightarrow{\delta^*} \check{H}^2(X; \mathbf{Z}) \xrightarrow{i^*} \check{H}^2(X; \mathcal{O}) \to \dots \quad (9.24)$$

Now in the particular case when X is $\mathbf{CP}^1$, we have seen that

$$\check{H}^1(X; \mathcal{O}) = 0 = \check{H}^2(X; \mathcal{O})$$

so that (9.24) reduces to

$$0 \to \check{H}^1(X; \mathcal{O}^*) \xrightarrow{\delta^*} \check{H}^2(X; \mathbf{Z}) \to 0$$

where exactness implies that δ^* is 1-1 and onto, that is, an isomorphism:

$$\check{H}^1(X; \mathcal{O}^*) \cong \check{H}^2(X; \mathbf{Z}). \quad (9.25)$$

The significance of this isomorphism is the following: a consideration of holomorphic line-bundles on $\mathbf{CP}^1$ (exercise 3g) shows that they are defined by 1-cocycles $C^1(X; \mathcal{O}^*)$. Further two line-bundles may be thought of as equivalent if the cocycles differ by a coboundary, since this just represents a coordinate transformation. Thus $\check{H}^1(X; \mathcal{O}^*)$ represents the equivalence classes of line-bundles. This will be true for any X. For the case considered here, one also knows that $\check{H}^2(X; \mathbf{Z}) = \mathbf{Z}$ (which is a particular instance of the general result that $\check{H}^n(X; \mathbf{Z}) = \mathbf{Z}$ for any n-dimensional real manifold

M). So (9.25) states that the group of equivalence classes of line-bundles over $\mathbf{CP}^1$ is isomorphic to the integers. This observation can be extended to say that every line-bundle over $\mathbf{CP}^1$ is equivalent to $\mathcal{O}(n)$ for some n.

In the next chapter, we return to the problem of solving the zero rest mass equations and find the theorem conjectured in (9.14).

Exercises 9

a) Show that the map $\pi : \mathcal{O} \to \mathbf{C}$ is a homeomorphism from $[f, D]$ to D.

b) Is there a sheaf $\mathcal{B}$ over $\mathbf{R}$ whose sections over the set $U \subset \mathbf{R}$ are bounded functions on U?

c) The procedure used in the definition of Čech cohomology defines a map

$$\check{H}^q(\{U_i\}; \mathcal{S}) \to \check{H}^q(X; \mathcal{S}).$$

Show that in the case $q = 1$ this is injective. (In other words, by taking the refinement one obtains a *larger* group.)

d) One particular fine resolution of the sheaf $\mathcal{O}$ on a complex manifold X which is of great importance is that leading to Dolbeault cohomology (Morrow and Kodaira 1971; Field 1982). To introduce this we need some definitions. A C^∞ 1-form α on, say, twistor space is said to be of type $(0, 1)$ iff it has the form $\alpha = A^{\bar{\beta}}(Z, \bar{Z})d\bar{Z}_\beta$. Likewise a 2-form is a type $(0, 2)$ form iff it has the form $A^{\bar{\alpha}\bar{\beta}}(Z, \bar{Z})d\bar{Z}_\alpha \wedge d\bar{Z}_\beta$. In general a (p, q) form has p dZ's and q $d\bar{Z}$'s. We write $\mathcal{E}^{p,q}$ for the sheaf of germs of (p, q) forms. Now define

$$\bar{\partial} = d\bar{Z}_\alpha \wedge \frac{\partial}{\partial \bar{Z}_\alpha} : \mathcal{E}^{p,q} \to \mathcal{E}^{p,q+1}$$

which is the anti-holomorphic version of the exterior derivative d. Show that

$$d = \partial + \bar{\partial}; \ \partial^2 = 0 = \bar{\partial}^2 = \partial\bar{\partial} + \bar{\partial}\partial.$$

Thus there is a complex

$$\cdots \xrightarrow{\bar{\partial}} \mathcal{E}^{p,q} \xrightarrow{\bar{\partial}} \mathcal{E}^{p,q+1} \xrightarrow{\bar{\partial}} \cdots.$$

The fact that this is exact is the Dolbeault lemma (Field 1982).

Further these sheaves are fine since any cover $\{U_i\}$ of X has a partition of unity consisting of C^∞ functions $\{p_i\}$ subordinate to it. Deduce that

$$0 \to \mathcal{O} \xrightarrow{i} \mathcal{E}^{0,0} \xrightarrow{\bar{\partial}} \mathcal{E}^{0,1} \xrightarrow{\bar{\partial}} \cdots$$

is a fine resolution of $\mathcal{O}$.

The cohomology groups

$$H^p = \frac{\ker \overline{\partial} : \mathcal{E}^{0,p} \to \mathcal{E}^{0,p+1}}{\mathrm{im}\ \overline{\partial} : \mathcal{E}^{0,p-1} \to \mathcal{E}^{0,p}}$$

obtained in this way are known as *Dolbeault* cohomology groups. In these terms a p-cohomology element is represented by a globally defined C^∞ $\overline{\partial}$-closed $(0,p)$-form rather than a set of holomorphic functions defined on $p + 1$-fold intersections. As a further exercise, can you see how these two representations are related for $H^1(\mathbf{CP}^1; \mathcal{O}(-2))$?

e) Fill in the gaps in the proof of Theorem 2.

f) If V and W are open sets in a complex manifold X, consider the sequence

$$0 \to C^p(V \cup W; \mathcal{S}) \overset{\sigma}{\to} C^p(V; \mathcal{S}) \oplus C^p(W; \mathcal{S}) \overset{\tau}{\to} C^p(V \cap W; \mathcal{S}) \to 0$$

where

$$\begin{aligned} \sigma(f) &= (\rho_V f, \rho_W f) \\ \tau(g, h) &= \rho_W g - \rho_V h. \end{aligned}$$

Show that this sequence is exact. The resulting long exact sequence is the *Mayer–Vietoris* sequence:

$$0 \to H^0(V \cup W; \mathcal{S}) \overset{\sigma^*}{\to} H^0(V; \mathcal{S}) \oplus H^0(W; \mathcal{S}) \overset{\tau^*}{\to} H^0(V \cap W; \mathcal{S}) \overset{\delta^*}{\to}$$
$$H^1(V \cup W; \mathcal{S}) \to \cdots .$$

Note that $\delta^* : H^0(V \cap W; \mathcal{S}) \to H^1(V \cup W; \mathcal{S})$ is an instance of the Čech map:

$$\check{C} : H^0(U_0 \cap U_1 \cap \ldots \cap U_p; \mathcal{S}) \to H^p(U_0 \cup U_1 \cup \ldots \cup U_p; \mathcal{S})$$

which interprets a function on an intersection as a representative cocycle.

Chapter 10

Solving the Zero Rest Mass Equations II

With the machinery of sheaf cohomology we now set out to demonstrate the generality of the contour integral formulae of chapter 8. We wish to prove two theorems:

1. $\check{H}^1(\mathbf{PT}^+; \mathcal{O}(-n-2)) \cong$ {zero rest mass fields $\varphi_{A' \ldots B'}(x)$ of helicity n holomorphic on $\mathbf{CM}^+$}. $\qquad$ (10.1)

2. $\check{H}^1(\mathbf{PT}^+; \mathcal{O}(n-2)) \cong$ {zero rest mass fields $\varphi_{A \ldots B}(x)$ of helicity $-n$ holomorphic on $\mathbf{CM}^+$}. $\qquad$ (10.2)

These correspond to (8.8) and (8.9).

The method of proof is to find a suitable short exact sequence of sheaves and deduce the theorems from the corresponding long exact sequence of cohomology. The right hand sides in (10.1) and (10.2) are globally defined objects, that is they have the character of $\check{H}^0$'s on $\mathbf{CM}^+$, while the left hand sides are $\check{H}^1$'s on $\mathbf{PT}^+$. Accordingly we consider sheaves on an intermediate object, the primed spin bundle, which is traditionally referred to as $\mathbf{F}$ in this context.

We have the double fibration (Eastwood et al. 1981; Wells 1981; see exercise 7f) in figure 10.1. Here ν is the usual projection; the inverse image of a point in $\mathbf{PT}$ under μ is the whole α-plane in $\mathbf{F}$. In particular, a function $f(x^a, \pi_{A'})$ on $\mathbf{F}$ pushes down to a function on $\mathbf{PT}$ if it is constant on α-planes or equivalently if

$$\pi^{A'} \nabla_{AA'} f = 0. \qquad (10.3)$$

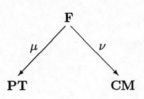

$$\nu \; : \; (x^a, \pi_{A'}) \to x^a$$
$$\mu \; : \; (x^a, \pi_{A'}) \to (ix^{AA'}\pi_{A'}, \pi_{A'}),$$

Figure 10.1.

We may restrict (10.1) to the future tube $\mathbf{CM}^+$ and corresponding regions $\mathbf{F}^+$ and $\mathbf{PT}^+$. We define $\mathcal{Z}'_n(m)$ to be the sheaf of germs of symmetric n-index primed spinor fields $\varphi_{A'\ldots B'}(x, \pi)$ holomorphic on $\mathbf{F}^+$, homogeneous of degree m in $\pi_{A'}$ and satisfying the zero rest mass equation:

$$\nabla_A{}^{A'}\varphi_{A'\ldots B'} = 0. \qquad (10.4)$$

Consider the map

$$\pi^{A'} : \mathcal{Z}'_{n+1}(m-1) \;\; \to \;\; \mathcal{Z}'_n(m)$$
$$\varphi_{A'B'\ldots C'} \;\; \to \;\; \pi^{A'}\varphi_{A'B'\ldots C'}. \qquad (10.5)$$

It is a straightforward exercise to see that this is onto, so we may consider the short exact sequence

$$0 \to \mathcal{T} \xrightarrow{i} \mathcal{Z}'_{n+1}(-1) \xrightarrow{\pi^{A'}} \mathcal{Z}'_n(0) \to 0 \qquad (10.6)$$

where $\mathcal{T}$ is just the kernel of the map $\pi^{A'}$. If $\psi_{A'\ldots B'} \in \mathcal{T}$ then this means that $\psi_{A'\ldots B'}\pi^{A'} = 0$ so that $\psi_{A'\ldots B'} = \pi_{A'}\ldots\pi_{B'}f(x, \pi)$. Further the zero rest mass equation (10.4) on $\psi_{A'\ldots B'}$ requires that $f(x, \pi)$ satisfy (10.3). From the homogeneities chosen in (10.6), f is therefore a twistor function homogeneous of degree $-n - 2$. We may write (10.6) as

$$0 \to \mathcal{T}(-n-2) \xrightarrow{\pi_{A'}\ldots\pi_{B'}} \mathcal{Z}'_{n+1}(-1) \xrightarrow{\pi^{A'}} \mathcal{Z}'_n(0) \to 0.$$

The corresponding long exact sequence of cohomology includes the section

$$\cdots \to \check{H}^0(\mathbf{F}^+; \mathcal{Z}'_{n+1}(-1)) \to \check{H}^0(\mathbf{F}^+; \mathcal{Z}'_n(0)) \to \check{H}^1(\mathbf{F}^+; \mathcal{T}(-n-2)) \to$$
$$\check{H}^1(\mathbf{F}^+; \mathcal{Z}'_{n+1}(-1)) \to \cdots$$

The first term is global sections of $\mathcal{Z}'_{n+1}(-1)$ over $\mathbf{F}^+$. For fixed x^a, such a section would give a global section homogeneous of degree -1 on $\mathbf{P}^1$, and there are no such sections. Thus $\check{H}^0(\mathbf{F}^+; \mathcal{Z}'_{n+1}(-1)) = 0$.

A similar argument, slightly more technical, shows that

$$\check{H}^1(\mathbf{F}^+; \mathcal{Z}'_{n+1}(-1)) = 0$$

so that we are left with an isomorphism

$$\check{H}^0(\mathbf{F}^+; \mathcal{Z}'_n(0)) \overset{\delta^*}{\cong} \check{H}^1(\mathbf{F}^+; \mathcal{T}(n-2)) \tag{10.7}$$

The term on the left represents solutions of the zero rest mass equations on $\mathbf{F}^+$, homogeneous of degree zero in $\pi_{A'}$. They must therefore be independent of $\pi_{A'}$, and defined on $\mathbf{CM}^+$. The term on the right is a cohomology group on $\mathbf{F}^+$ but the coefficients are *twistor* functions, so that they don't notice that they aren't defined on $\mathbf{PT}^+$, i.e.

$$\check{H}^1(\mathbf{F}^+; \mathcal{T}(-n-2)) \cong \check{H}^1(\mathbf{PT}^+; \mathcal{O}(-n-2)). \tag{10.8}$$

With these observations, (10.7) is the required isomorphism (10.1).

(Evidently there are some non-trivial details, which we have glossed over here, to do with relating cohomology on different spaces and to do with the treatment of infinity in $\mathbf{CM}^+$. These and similar details omitted below may be found in Eastwood et al. (1981).)

The map δ^{*-1} which, given an element of $\check{H}^1(\mathbf{PT}^+; \mathcal{O}(-n-2))$, produces a z.r.m. field, may be described in terms of the following diagram of cochain groups:

$$
\begin{array}{ccccccccc}
& & \downarrow & & \downarrow & & \downarrow & & \\
0 & \to & C^0(\mathcal{T}(-n-2)) & \to & C^0(\mathcal{Z}'_{n+1}(-1)) & \to & C^0(\mathcal{Z}'_n(0)) & \to & 0 \\
& & \downarrow & & \downarrow & & \downarrow & & \\
0 & \to & C^1(\mathcal{T}(-n-2)) & \to & C^1(\mathcal{Z}'_{n+1}(-1)) & \to & C^1(\mathcal{Z}'_n(0)) & \to & 0 \\
& & \downarrow & & \downarrow & & \downarrow & &
\end{array}
$$

Suppose f_{ij} defines an element of $\check{H}^1(\mathbf{PT}^+; \mathcal{O}(-n-2))$ with respect to some chosen cover. Then

$$\pi_{A'} \ldots \pi_{B'} \pi_{C'} f_{ij} \in C^1(\mathcal{Z}'_{n+1}(-1))$$

and

$$\delta(\pi_{A'} \ldots \pi_{C'} f_{ij}) = \pi_{A'} \ldots \pi_{C'} \delta f_{ij} = 0.$$

But $\check{H}^1(\mathcal{Z}'_{n+1}(-1)) = 0$ so that

$$\pi_{A'} \ldots \pi_{B'} \pi_{C'} f_{ij} = \rho_{[i} \psi_{j]A' \ldots B'C'} \tag{10.9}$$

for $\psi_{jA'...B'C'} \in C^0(\mathcal{Z}'_{n+1}(-1))$. Define

$$\varphi_{jA'...B'} = \psi_{jA'...B'C'}\pi^{C'} \in H^0(\mathcal{Z}'_n(0)) \tag{10.10}$$

then by (10.8)

$$\rho_{[i}\varphi_{j]A'...B'} = 0$$

whence

$$\varphi_{jA'...B'} = \rho_j \varphi_{A'...B'} \tag{10.11}$$

and $\varphi_{A'...B'}$ is the required z.r.m. field. To see that it satisfies the z.r.m. field equations we have from (10.10)

$$\rho_{[i}\nabla^{AC'}\psi_{j]A'...B'C'} = 0$$

so that $\nabla^{AC'}\psi_{jA'...B'C'}$ is globally defined, homogeneous of degree -1, and so vanishes. Thus by (10.10) and (10.11)

$$\nabla^{A'}_A \varphi_{A'...B'} = 0.$$

The contour integration of chapter 8 is done in the splitting implicit in (10.9) (recall equation 9.8).

So much for (10.1). The negative helicity result (10.2) is different in that it involves potentials. We start with the short exact sequence of sheaves on $\mathbf{F}^+$:

$$0 \to \mathcal{T}(n) \to \mathcal{K}(n) \xrightarrow{D_A} \mathcal{Q}_A(n+1) \to 0 \tag{10.12}$$

where

$\mathcal{K}(n)$ is the sheaf of germs of holomorphic functions on $\mathbf{F}^+$ homogeneous of degree n in $\pi_{A'}$;

D_A is $\pi^{A'}\nabla_{AA'}$ and $\mathcal{T}(n)$ is the kernel of this map, i.e. $\mathcal{T}(n)$ represents homogeneity n twistor functions;

$\mathcal{Q}_A(n+1)$ is the sheaf of germs of spinor-valued functions $\psi_A(x,\pi)$ homogeneous of degree $(n+1)$ in $\pi_{A'}$ and satisfying $D^A\psi_A = 0$.

The only part of exactness which isn't obvious is that D_A is onto $\mathcal{Q}_A(n+1)$, and this is quite straightforward.

The corresponding long exact sequence of cohomology includes the section

$$0 \to \check{H}^0(\mathbf{F}^+; \mathcal{T}(n)) \to \check{H}^0(\mathbf{F}^+; \mathcal{K}(n)) \to \check{H}^0(\mathbf{F}^+; \mathcal{Q}_A(n+1))$$
$$\to \check{H}^1(\mathbf{F}^+; \mathcal{T}(n)) \to \check{H}^1(\mathbf{F}^+; \mathcal{K}(n)) \to \ldots \tag{10.13}$$

and we must identify these various groups:

$\check{H}^0(\mathbf{F}^+; \mathcal{T}(n))$ is the space of global twistor functions on $\mathbf{PT}^+$ homogeneous of degree n. These must be of the form of polynomials $\mu = \mu_{A'...B'}\pi^{A'}\ldots\pi^{B'}$ satisfying

$$\pi^{C'}\nabla_{CC'}\mu = \pi^{A'}\ldots\pi^{B'}\pi^{C'}\nabla_{C(C'}\mu_{A'...B')} = 0. \qquad (10.14)$$

$\check{H}^0(\mathbf{F}^+; \mathcal{K}(n))$ is the space of polynomials like μ but without (10.14), say $\lambda = \lambda_{A'...B'}\pi^{A'}\ldots\pi^{B'}$.

$\check{H}^0(\mathbf{F}^+; \mathcal{Q}_A(n+1))$ is the space of spinor fields $\psi_A = \psi_{AA'...C'}\pi^{A'}\ldots\pi^{C'}$ with $(n+1)$ primed indices and satisfying

$$D^A\psi_A = \pi^{D'}\pi^{A'}\ldots\pi^{C'}\nabla_{(D'}{}^A\psi_{A'...C')A} = 0 \qquad (10.15)$$

We shall write these spaces respectively as T_n, Λ_n and Ψ_n. Next, by similar arguments to those leading to (10.8), we have $\check{H}^1(\mathbf{F}^+; \mathcal{T}(n)) = \check{H}^1(\mathbf{PT}^+; \mathcal{O}(n))$. Finally, and this we must ask the reader to take on trust, $\check{H}^1(\mathbf{F}^+; \mathcal{K}(n)) = 0$.

(10.13) becomes

$$0 \to T_n \xrightarrow{i} \Lambda_n \xrightarrow{\sigma} \Psi_n \xrightarrow{\delta^*} \check{H}^1(\mathbf{PT}^+; \mathcal{O}(n)) \to 0 \qquad (10.16)$$

where i is inclusion, σ is the map

$$\lambda_{A'...B'} \to \nabla_{A(A'}\lambda_{B'...C')}, \qquad (10.17)$$

and δ^* is the connecting homomorphism (which we shall describe explicitly in a moment).

Now how does this help us to identify $\check{H}^1(\mathbf{PT}^+; \mathcal{O}(n))$? The answer is in terms of 'potentials modulo gauge transformations'. Given a field $\psi_{AA'...B'C'}$ satisfying (10.15):

$$\nabla^A{}_{(D'}\psi_{A'...C')A} = 0$$

we define

$$\varphi_{AB...D} = \nabla^{B'}_{(B...}\nabla^{D'}_D\psi_{A)B'...D'}. \qquad (10.18)$$

Then this is a helicity $-\frac{1}{2}(n+2)$ z.r.m. field, i.e. it has $(n+2)$ indices and

$$\nabla^{AA'}\varphi_{AB...D} = 0. \qquad (10.19)$$

Deducing (10.19) from (10.18) with the aid of (10.15) is a straightforward calculation (and a good exercise in spinor calculus!). The transformation

$$\psi_{AA'...B'C'} \to \psi_{AA'...B'C'} + \nabla_{A(A'}\lambda_{B'...C')} \qquad (10.20)$$

is a gauge transformation, taking solutions of (10.15) to solutions of (10.15) but leaving unchanged $\varphi_{A...D}$ as defined by (10.18). Finally, if $\lambda_{B'...C'}$ is a solution of the twistor equation (10.14), then (10.20) actually leaves $\psi_{AA'...C'}$ invariant.

Thus there is an exact sequence

$$0 \to T_n \xrightarrow{i} \Lambda_n \xrightarrow{\sigma} \Psi_n \xrightarrow{\nu} \Phi_{n+2} \to 0 \qquad (10.21)$$

where Φ_{n+2} is the group of helicity $-\frac{1}{2}(n+2)$ z.r.m. fields on $\mathbf{CM}^+$ and ν is the map (10.18). The only part of exactness not checked is that ν is onto. This is readily done, and then a comparison of (10.21) with(10.11) yields the claimed isomorphism (10.2). (For Maxwell fields, the sequence of fields introduced here is more familiar. The z.r.m. field is φ_{AB}, the potential is $\psi_{AA'}$ satisfying

$$\nabla^A{}_{(A'}\psi_{B')A} = 0$$

and the gauge transformation is just

$$\psi_{AA'} \to \psi_{AA'} + \nabla_{AA'}\Lambda.$$

Finally $T_0 = \mathbf{C}$.)

The map δ^* can be described quite explicitly in this case and provides a partial answer to the problem of defining an 'inverse twistor function'; i.e. given the field, how do you find the corresponding $\check{H}^1$? Suppose the field is $\varphi_{A...D}$ and that $\psi_{AA'...C'}$ is a potential for it. For a given twistor $Z_i^\alpha \in \mathbf{PT}^+$ define a neighbourhood $U_i \subset \mathbf{PT}^+$ as the union of all the lines through Z_i^α which lie entirely in $\mathbf{PT}^+$.

Now given two such neighbourhoods U_i and U_j we define $f_{ij}(Z^\alpha)$ by an integral: if $Z^\alpha \in U_i \cap U_j$ then the α-plane defined by Z^α meets the α-planes defined by Z_i^α and Z_j^α at P_i and P_j in $\mathbf{CM}^+$ (see figure 10.2). Define

$$f_{ij}(Z^\alpha) = \int_\Gamma \psi_{AA'B'...C'}\pi^{B'}\cdots\pi^{C'} dx^{AA'} \qquad (10.22)$$

where $Z^\alpha = (\omega^A, \pi_{A'})$ and the integral is along a path Γ lying in the α-plane Z^α and joining its intersections P_i, P_j with the α-planes Z_i^α and Z_j^α. Because of (10.15), the integral is independent of the path chosen so that (10.22) *is* a function of Z. By the same token, this defines a *cocycle* in that

$$f_{ij} + f_{jk} + f_{ki} = 0$$

since this represents an integral round a closed contour in the α-plane.

To see that this is the *right* cocycle, we return to the question of extracting the field from a given cocycle.

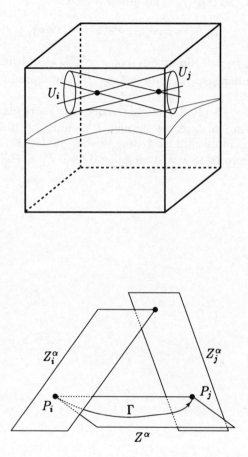

Figure 10.2. The inverse twistor function.

Given a cocycle $f_{ij} \in \check{H}(\mathbf{PT}^+; \mathcal{O}(n))$, first restrict it to a line in $\mathbf{PT}^+$ to get $\rho_x f_{ij} \in \check{H}(\mathbf{P}^1; \mathcal{O}(n))$. This group is zero so

$$\rho_x f_{ij} = \rho_{[i} \psi_{j]}, \ \psi_j \in C^0(\mathbf{P}^1; \mathcal{O}(n)) \tag{10.23}$$

also $\pi^{A'} \nabla_{AA'} \rho_x f_{ij} = 0$. So $\pi^{A'} \nabla_{AA'} \psi_i = \pi^{A'} \nabla_{AA'} \psi_j$ defines a polynomial of degree $(n+1)$ in $\pi_{A'}$, $\psi_{AA'...B'}(x) \pi^{A'} \cdots \pi^{B'}$, which produces the potential $\psi_{AA'...B'}(x)$.

With the cocycle (10.22), if we restrict to a particular point P, then we are considering all α-planes through P as in figure 10.2. The splitting (10.23) is now accomplished by writing the integral (10.22) as the difference of an integral from P_i to P and an integral from P_j to P. The rest follows immediately.

Chapter 11

The Twisted Photon and Yang–Mills Constructions

Up to now we have encountered twistor functions (or more accurately, cohomology classes) in a purely *passive* role: a z.r.m. field results from doing something to a twistor function. In the next two chapters, the twistor functions have a more *active* role in that they are used to *change* twistor space.

The first example is the *twisted photon* (Hughston and Ward 1979) where a function $f_0(Z^\alpha)$ homogeneous of degree zero is used to define a line bundle which encodes the information of a Maxwell field.

We recall that the space **PT** is the space of α-planes in **CM**. Given one such α-plane, the extra information in **T** represents the choice of a constant $\pi_{A'}$-spinor associated to the α-plane. Equivalently, the fibre of **T** thought of as a line bundle over **PT** consists of the solutions $\pi_{A'}$ of the equation

$$X^b \nabla_b \pi_{A'} = 0 \tag{11.1}$$

where X^b denotes the tangent vectors to the α-plane,

$$X^b = \lambda^B \pi^{B'}, \text{ for arbitrary } \lambda^B.$$

Writing this equation as

$$\pi^{B'} \nabla_{BB'} \pi_{A'} = 0$$

we consider ways of modifying it so as to construct different bundles over **PT**. One possibility is to consider

$$\pi^{B'} D_{BB'} \pi_{A'} \equiv \pi^{B'} (\nabla_{BB'} - ieA_{BB'}) \pi_{A'} = 0 \tag{11.2}$$

99

for constant e and some vector $A_{BB'}$. Solutions will exist if the relevant curvature vanishes, i.e. we must require

$$\epsilon^{AB}\pi^{A'}D_{AA'}\pi^{B'}D_{BB'}\pi_{C'} = 0$$

which is just

$$\pi^{A'}\pi^{B'}\nabla_{AA'}A_{B'}{}^{A} = 0. \tag{11.3}$$

This is true for all $\pi^{A'}$ iff

$$\nabla_{A(A'}A_{B')}{}^{A} = 0. \tag{11.4}$$

Thus given $A_{AA'}$ satisfying (11.4), (11.2) is integrable on all α-planes and we have a holomorphic complex line bundle T over **PT** (or at least over the part of **PT** corresponding to the part of **CM** on which $A_{AA'}$ is defined) whose fibre is the one-dimensional vector space of solutions to (11.2). In fact, given (11.4) we can solve

$$\pi^{A'}D_{AA'}\xi_{B'} = 0 \tag{11.5}$$

on each α-plane, i.e. we don't have to consider only solutions proportional to $\pi_{A'}$. If we focus attention on one point p of **CM** and consider all the α-planes through it, then we can compare solutions of (11.5) on the different α-planes by comparing them at p. Thus the bundle T restricted to the line L_p in **PT** must be trivial.

Now from the discussion in chapter 10, if we define

$$\varphi_{AB} = \nabla_{A'(A}A_{B)}{}^{A'} \tag{11.6}$$

then φ_{AB} satisfies Maxwell's equations:

$$\nabla_{A'}{}^{A}\varphi_{AB} = 0 \tag{11.7}$$

as a consequence of (11.4). Conversely, any a.s.d. Maxwell field φ_{AB} has a potential $A_{AA'}$ satisfying (11.4) so we may enunciate a

Theorem

There is a one-to-one correspondence between anti-self-dual Maxwell fields on **CM**$^{+}$ *and complex line bundles on* **PT**$^{+}$ *which are trivial on each line* L_p.

Proof

The correspondence one way is shown above. We consider below the question of going the other way. First, a construction analogous to that of

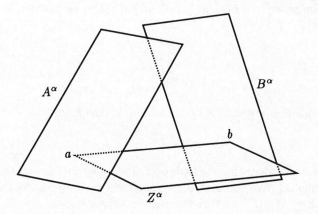

Figure 11.1. Constructing $\mathcal{T}$.

figure 10.1 provides an explicit coordinatisation: suppose A^α, B^α play the roles of Z_i^α and Z_j^α there (figure 11.1). Then for the solutions at p of the equation (11.5) on the α-plane Z^α we may start with $\pi_0^{A'}$ at a to obtain

$$\pi^{A'} = \pi_0^{A'} \exp(ie \int_a^p A_b dx^b).$$

or start with $\pi_0^{A'}$ at b to obtain

$$\hat{\pi}^{A'} = \pi_0^{A'} \exp(ie \int_b^p A_b dx^b)$$

The two different answers are related by

$$\begin{aligned}
\hat{\pi}^{A'} &= \pi^{A'} \exp(ie \int_b^a A_b dx^b) \\
&= \pi^{A'} \exp(ief_0(Z^\alpha)). \quad (11.8)
\end{aligned}$$

The twisted photon construction may therefore be described as follows: given a line bundle L over $\mathbf{PT}^+$ trivial on lines, trivialise L with respect to some cover $\{U_i\}$ of $\mathbf{PT}^+$. Then L is coordinatised with coordinates Z_i^α patched according to

$$Z_i^\alpha = Z_j^\alpha \exp(ief_{ij}(Z^\alpha)) \quad (11.9)$$

where $f_{ij}(Z^\alpha)$ defines a cocycle in $\check{H}^1(\mathbf{PT}^+; \mathcal{O}(0))$. To extract a field, we reverse the process leading to (11.8): given (11.9) we restrict to a line L_x and 'split' f

$$\rho_x f_{ij} = \rho_{[i}\psi_{j]}$$

then

$$\pi^{A'}\nabla_{AA'}\psi_i = \pi^{A'}\nabla_{AA'}\psi_j = A_{AA'}(x)\pi^{A'}$$

which yields the potential $A_{AA'}(x)$ and hence the field.

$\square$

There is an interesting example of charge integrality associated with the twisted photon. We consider the complement of a world line in $\mathbf{M}$. This has topology $\mathbf{R} \times S^2$ so that the corresponding region, say X, in $\mathbf{PT}$ is topologically $\mathbf{R} \times S^2 \times S^2$ where the second S^2 is the celestial sphere. On X consider the familiar short exact sequence of sheaves

$$0 \to \mathbf{Z} \to \mathcal{O} \to \mathcal{O}^* \to 0$$

and the portion of the corresponding long exact sequence of cohomology groups

$$H^1(X; \mathbf{Z}) \to H^1(X; \mathcal{O}) \to H^1(X; \mathcal{O}^*) \xrightarrow{\delta^*} H^2(X; \mathbf{Z}).$$

A topological calculation shows that $H^1(X; \mathbf{Z}) = 0$ while $H^2(X; \mathbf{Z}) = \mathbf{Z} \otimes \mathbf{Z}$ so we are left with

$$0 \to H^1(X; \mathcal{O}) \to H^1(X; \mathcal{O}^*) \xrightarrow{\delta^*} \mathbf{Z} \otimes \mathbf{Z}.$$

The first group $H^1(X; \mathcal{O})$ represents Maxwell fields which can be obtained by the contour integral formula (8.9), while the second group $H^1(X; \mathcal{O}^*)$ represents Maxwell fields which can be obtained as line bundles by the twisted photon construction. We claim that the map δ^* goes to zero in one of the $\mathbf{Z}$'s but goes to the value of the *charge* of the Maxwell field in the other. (This requires choosing representatives carefully and is discussed in Hughston and Ward (1979).) Consequently, Maxwell fields constructed as twisted photons have integral charge, while Maxwell fields constucted as contour integrals have *zero* charge (since the sequence is exact at $H^1(X; \mathcal{O}^*)$).

The twisted photon construction can be generalised in a natural way to produce solutions to the anti-self-dual Yang–Mills equation. This is rather more remarkable since these equations are non-linear. A Yang–Mills field is essentially a connection on a trivial $GL(n, \mathbf{C})$ bundle B over $\mathbf{M}$.

We may represent the connection by a matrix of potentials $A_{AA'i}{}^j$ where $i, j = 1, \ldots, n$. The covariant derivative is

$$D_{AA'i}{}^j = \delta_i{}^j \nabla_{AA'} - ie A_{AA'i}{}^j \qquad (11.10)$$

and curvature is obtained by commuting derivatives:

$$F_{abi}{}^j = \nabla_a A_{bi}{}^j - \nabla_b A_{ai}{}^j - ie(A_{ai}{}^k A_{bk}{}^j - A_{bi}{}^k A_{ak}{}^j). \qquad (11.11)$$

We may write this in a matrix notation omitting the Yang–Mills indices as

$$F_{ab} = 2\nabla_{[a} A_{b]} - ie[A_a, A_b] \qquad (11.12)$$

where the large brackets denote a matrix commutator. From (11.12) one obtains a Bianchi identity

$$\nabla_{[a} F_{bc]} + ie(F_{[ab} A_{c]} - A_{[a} F_{bc]}) = 0. \qquad (11.13)$$

Finally, the Yang–Mills field equations are

$$\nabla_{[a}{}^* F_{bc]} + ie({}^* F_{[ab} A_{c]} - A_{[a}{}^* F_{bc]}) = 0. \qquad (11.14)$$

In general, (11.14) represents a second-order equation on the potential matrix A_a. However, if the original choice of A_a produced a field F_{ab} which was self-dual or anti-self-dual then the field equations (11.14) are a consequence of the Bianchi identities (11.13). In this case the equation to be solved is first order in A_a, namely

$$^* F_{ab} = \pm i F_{ab}.$$

To find a twistorial approach to this problem we consider the analogue of (11.5). That is, on a particular α-plane specified by $\pi^{A'}$ we consider the equations

$$\pi^{A'} D_{AA'i}{}^j \varphi_j \equiv \pi^{A'}(\nabla_{AA'} \delta_i{}^j - ie A_{AA'i}{}^j) \varphi_j = 0. \qquad (11.15)$$

This equation will have n linearly independent solutions on the α-plane iff

$$\epsilon^{AB} \pi^{A'} \pi^{B'} D_{AA'i}{}^j D_{BB'j}{}^k = 0. \qquad (11.16)$$

This is the counterpart of (11.3) and implies

$$\nabla_{A(A'} A_{B')}{}^A{}_i{}^j - ie A_{A(A'i}{}^k A_{B')}{}^A{}_k{}^j = 0 \qquad (11.17)$$

or equivalently

$$^* F_{ab} = -i F_{ab}.$$

Given this condition of anti-self-duality on the field, one may construct a bundle $\mathcal{E}$ over some part of **PT** whose fibre above a particular α-plane is the vector space of solutions to (11.15). By the same argument as before, $\mathcal{E}$ is trivial when restricted to a line L_p in **PT** so that we may enunciate another

Theorem

There is a one-to-one correspondence between $GL(n, \mathbf{C})$ Yang–Mills fields on $\mathbf{CM}^+$ and rank n holomorphic vector bundles on $\mathbf{PT}^+$ which are trivial on each line, L_p.

Proof

The procedure for extracting the field from the bundle is precisely analogous to the previous method. If the transition matrix is $G(Z^\alpha)$ then when restricted to a line it must split but now as a product:

$$\rho_x G(Z^\alpha) = \hat{g}^{-1}(x, \pi) g(x, \pi)$$

or

$$\hat{g} G = g.$$

Now $\pi^{A'} \nabla_{AA'} \hat{g}.G = \pi^{A'} \nabla_{AA'} g$ since $\pi^{A'} \nabla_{AA'} G = 0$. Thus

$$
\begin{aligned}
\pi^{A'} \nabla_{AA'} \hat{g}.\hat{g}^{-1} &= \pi^{A'} \nabla_{AA'} g.g^{-1} \\
&= ie A_{AA'} \pi^{A'}
\end{aligned}
\tag{11.18}
$$

where $A_{AA'}$ is the matrix of potentials. Applying $\pi^{B'} \nabla^A{}_{B'}$ to (11.18) immediately gives (11.17).

$\square$

For subgroups of $GL(n, \mathbf{C})$, notably $SU(n)$, extra structures are required on the bundles $\mathcal{E}$ but we shall not consider this further. Instead we refer to the original papers of Ward (1977) and Atiyah and Ward (1977) and the book of Ward and Wells (1990) where also specific examples are given. In conclusion, we remark that what allows the constructon to work is that an anti-self-dual connection is *flat* on α-planes. In the next chapter we shall see another instance of this.

Chapter 12

The Non-Linear Graviton

Twistor theory as discussed so far has been concerned solely with flat space, or at most with conformally flat space. There are various ways of attempting to generalise different parts of twistor theory in the presence of conformal curvature, and these meet with varying degrees of success (Penrose 1976; Penrose and Ward 1980; Ward and Wells 1990; Bailey and Baston 1990). In this chapter we wish to consider the problem of finding α-planes in a space-time $\mathbf{M}$ with conformal curvature. Naturally we shall need to complexify $\mathbf{M}$ since an α-plane is a complex surface, so we need to suppose $\mathbf{M}$ is analytic.

Two tangents X^a, Y^a to an α-plane necessarily have the form

$$X^a = \lambda^A \pi^{A'}, \ \ Y^a = \mu^A \pi^{A'} \tag{12.1}$$

for some $\pi^{A'}$. To be surface-forming, the commutator of X^a and Y^a must be a linear combination of X^a and Y^a:

$$X^b \nabla_b Y^a - Y^b \nabla_b X^a = \alpha X^a + \beta Y^a. \tag{12.2}$$

The RHS of (12.2) has the form $\nu^A \pi^{A'}$ for some ν^A so that (12.2) is equivalent to

$$\pi^{A'}(X^b \nabla_b Y_a - Y^b \nabla_b X_a) = 0.$$

Substituting from (12.1) we are led to

$$\pi^{A'} \pi^{B'} \nabla_{BB'} \pi_{A'} = 0 \tag{12.3}$$

which we recognise as the geodesic shear-free condition of chapter 6. Applying $\pi^{C'} \nabla_{C'}{}^B$ to (12.3) we are led to

$$\overline{\Psi}_{A'B'C'D'} \pi^{A'} \pi^{B'} \pi^{C'} \pi^{D'} = 0.$$

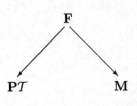

Figure 12.1.

(This could have been deduced from the Sachs equations (6.13,6.14.)

Thus if at a point p in **M** we want an α-plane for each primed spinor $\pi_{A'}$, then the primed Weyl spinor $\overline{\Psi}_{A'B'C'D'}$ must vanish. For a real space-time this necessarily entails that Ψ_{ABCD} vanish and **M** be conformally flat. However, if we consider instead *complex* space-times **M** then we lose the conjugation of spinors, and the primed and unprimed Weyl spinors become independent curvature quantities. Now we *can* set one part equal to zero. The Weyl tensor is required to be a.s.d. in that

$$^{*}C_{abcd} = \frac{1}{2}\epsilon_{ab}{}^{pq}C_{pqcd} = -iC_{abcd}. \tag{12.4}$$

(In a four-dimensional manifold of some other signature, either $(++++)$ or $(++--)$, the duality operation has *real* eigenvalues and one can have *real* solutions with a.s.d. Weyl tensor. The present discussion will still cover the *complexification* of these solutions.)

Thus given (12.4) we can define a three-dimensional complex manifold **PT**, the space of α-planes in **M**. The equation

$$\pi^{A'}\nabla_{AA'}\xi_{B'} = 0$$

has solutions on the α-plane defined by $\pi_{A'}$, so we can also define T. The extra information in T is the choice of scale for the spinor $\pi_{A'}$ on each α-plane.

Another way to define T and **PT** is in terms of the primed spin bundle **F** of **M**, mimicking the double fibration of chapter 10 (see figure 12.1). Consider the distribution defined on **F** by the two vector fields $\pi^{A'}\nabla_{AA'}$. This distribution is integrable iff (12.4) holds. The space of leaves is **PT**, with T as above. The vector field $\Upsilon = \pi_{A'}\frac{\partial}{\partial \pi_{A'}}$ on **F** projects to the homogeneity operator on T, and **PT** is the space of integrable curves of Υ.

A point p of **M** is represented in **PT** by the set L_p of α-planes through p. As in the flat case, this is a projective line $\mathbf{P}^1$, and so **PT** has a special four-parameter family of compact holomorphic curves in it. Furthermore, null separation in $\mathbf{M}$ is mirrored by intersection of the corresponding curves in $\mathbf{P}\mathcal{T}$. The fact that the conformal structure in $\mathbf{M}$ is quadratic must be reflected in $\mathbf{P}\mathcal{T}$ by some restriction on these curves; specifically, the condition for two nearby curves to intersect must be a quadratic condition. To make this condition precise, we must define the *normal bundle* N of a sub-manifold Y in a manifold X (Hirzebruch 1966; Ward and Wells 1990). At a point $p \in Y$ the fibre N_p of N is

$$N_p = T_p X / T_p Y$$

i.e. N_p consists of vectors modulo vectors tangent to Y.

If Y is a $\mathbf{P}^1$ corresponding to a line L_p, then N is a rank two vector bundle over $\mathbf{P}^1$ and so, by a theorem of Grothendieck (1957), is a sum of line bundles:

$$N = \mathcal{O}(m) \oplus \mathcal{O}(n). \qquad (12.5)$$

(The vector bundle $\mathcal{O}(m) \oplus \mathcal{O}(n)$ has transition matrices $\text{diag}(f, g)$ where f, g are the transition functions for $\mathcal{O}(m), \mathcal{O}(n)$ respectively.)

An 'infinitesimally neighbouring' line L_q to L_p is more accurately a section of N, the normal bundle to L_p. By (12.5), L_q is defined by a pair of sections, one of $\mathcal{O}(m)$ and one of $\mathcal{O}(n)$. There is supposed to be a four-parameter family of nearby lines and by table 9.1

$$\dim \check{H}^0(\mathbf{P}^1; \mathcal{O}(m)) = \begin{cases} m+1 & m > -1 \\ 0 & m \leq -1 \end{cases}$$

so that m and n in (12.5) are severely constrained! Clearly

$$\begin{aligned} m+n = 2 & \quad \text{if } m, n > -1 \\ n = 3 & \quad \text{if } m \leq -1 \end{aligned}$$

so that the only possibilities for (m, n), if we order them with $m \leq n$, are $(-1, 3)$, $(0, 2)$ and $(1, 1)$. In the first case, all sections are of the form $(0, \varphi^{A'B'C'} \pi_{A'} \pi_{B'} \pi_{C'})$ where $\pi_{A'}$ is the coordinate on $\mathbf{P}^1$ and so all have 3 zeroes. In the second case, all sections are of the form $(C, \varphi^{A'B'} \pi_{A'} \pi_{B'})$ and so have either no zeroes (if $C \neq 0$) or 2 zeroes. In the third case, the sections are of the form $(\alpha^{A'} \pi_{A'}, \beta^{A'} \pi_{A'})$. Such a section has a zero iff the zeroes of the two parts coincide, i.e. iff $\alpha^{A'} \beta_{A'} = 0$. This is the required quadratic condition and so the restriction that we are looking for is that the four-parameter family of compact holomorphic curves in $\mathbf{P}\mathcal{T}$ must have normal bundle $N = \mathcal{O}(1) \oplus \mathcal{O}(1)$.

A twistor space $\mathbf{P}\mathcal{T}$ with these properties corresponds to a space-time $\mathbf{M}$ with a.s.d. Weyl tensor. In fact $\mathbf{P}\mathcal{T}$ corresponds to the whole class of space-times conformal to **M**, since the construction so far is conformally invariant.

We next suppose that **M** satisfies the vacuum field equations, i.e. that

$$\Phi_{ABA'B'} = 0 = \Lambda \tag{12.6}$$

as well as $\overline{\Psi}_{A'B'C'D'}$. The extra structure that this determines on **P**$\mathcal{T}$ arises from the fact that the equation

$$\nabla_{AA'}\pi_{B'} = 0$$

now has solutions. That is, we have a global parallelism of primed spinors; we can compare primed spinors not just all over each α-plane, but from α-plane to α-plane, and we can determine when two separate α-planes have the same $\pi_{A'}$. This gives **P**$\mathcal{T}$ the structure of a fibration with projection π over the projective primed spin-space **P**S, and $\mathcal{T}$ the same over the non-projective primed spin-space, S. Also, the base-space has a preferred choice of $\epsilon^{A'B'}$ inherited from the space-time **M**. Equivalently, there is a two-form $\tau = \epsilon^{A'B'} d\pi_{A'} \wedge d\pi_{B'}$ on $\mathcal{T}$.

The final extra piece of structure, which encodes the other piece ϵ_{AB} of the space-time metric, is a two-form μ on each fibre of the projection π. To define μ in space-time terms, suppose we are given a particular α-plane z with a particular $\pi_{A'}$, and a pair of vectors X, Y tangent to the fibre of $\mathcal{T}$ containing z. X and Y determine connecting vectors X^a and Y^a as fields on the α-plane z, connecting it to two 'nearby' α-planes with the same $\pi_{A'}$. Now form

$$\mu(X, Y) = \epsilon_{AB} X^{AA'} Y^{BB'} \pi_{A'} \pi_{B'}. \tag{12.7}$$

We claim that this is independent of the point on the α-plane z at which it is evaluated and independent of the particular choices of connecting vectors X^a and Y^a. This is easily verified, so that (12.7) does define a two-form on the fibres.

(Alternatively, μ may be defined on the primed spin-bundle **F** as

$$\mu = \frac{1}{2}\epsilon_{AB}\pi_{A'}\pi_{B'}dx^{AA'} \wedge dx^{BB'}$$

and projected down to $\mathcal{T}$.)

In summary, we have constructed from a solution **M** of the vacuum equations with a.s.d. Weyl curvature a four-dimensional complex manifold $\mathcal{T}$ with

i) a projection π to primed spin-space;

ii) a homogeneity operator Υ;

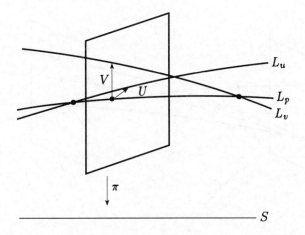

Figure 12.2. The arrangement for defining the metric.

iii) forms τ on the base and μ on each fibre with $\mathcal{L}_\Upsilon \tau = 2\tau$; $\mathcal{L}_\Upsilon \mu = 2\mu$;

iv) a four-parameter family of holomorphic curves which in $\mathbf{P}\mathcal{T}$ are compact and have normal bundle $\mathcal{O}(1) \oplus \mathcal{O}(1)$.

We must show that conversely such a $\mathcal{T}$ produces an $\mathbf{M}$, and then how to set about constructing such manifolds.

For the first part we need a direct construction of the metric and connection of $\mathbf{M}$ from $\mathcal{T}$ and $\mathbf{P}\mathcal{T}$. The conformal metric is, we know, defined by intersection, and condition (iv) ensures that it is quadratic. To fix the conformal factor of the metric at each point p it is sufficient to define the inner product of two null vectors, say U^a and V^a, at p. In $\mathbf{P}\mathcal{T}$ we suppose U^a and V^a correspond to lines L_u, L_v infinitesimally neighbouring to L_p. On each fibre of the projection π, L_u and L_v define vectors U and V. Further, since U^a is a null vector, U vanishes at one value of $\pi_{A'}$, say $\pi_{A'} = \alpha_{A'}$, and similarly V vanishes at another, say $\pi_{A'} = \beta_{A'}$ (figure 12.2). At a general value of $\pi_{A'}$ we may form $\mu(U, V)$. This is a function on the Riemann sphere L_p homogeneous of degree 2 and with precisely two zeroes, one each at $\alpha_{A'}$ and $\beta_{B'}$. We consider the expression

$$g(U, V) = \frac{(\alpha^{A'} \beta_{A'}) \mu(U, V)}{(\alpha^{B'} \pi_{B'})(\beta^{C'} \pi_{C'})}.\tag{12.8}$$

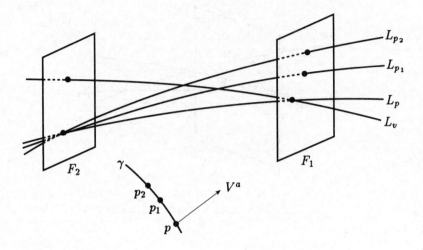

Figure 12.3. The connection.

The RHS is homogeneous of degree zero in $\pi_{A'}$ and has no zeroes. It is therefore *constant*, i.e. independent of $\pi_{A'}$, and we take this as the definition of $g_{ab}U^aV^b$. (Note that in order to construct (12.8) we have used both the forms τ and μ of condition (iii).)

A point in $\mathbf{PT}$ defines a totally null two-plane (that is, an α-plane in $\mathbf{M}$) by all the lines L_p through it. A *one*-parameter family of lines through a point z in $\mathbf{PT}$ therefore defines a curve lying in an α-plane. If we define a geometrical procedure for parallelly propagating a null vector along such a curve this is sufficient to define a connection on $\mathbf{M}$.

Suppose we have the curve γ through a point p and a null vector V^a at p. In $\mathbf{PT}$, two fibres are picked out, one F_1 corresponding to V^a and one F_2 corresponding to γ (figure 12.3).

To define vectors along γ corresponding to the parallel propagation of V^a, we take the lines joining the point of intersection of L_v with F_2 to the points of intersection of $L_{p_1}, L_{p_2}, \ldots$ with F_1. This is evidently a linear operation on vectors at p. Further, if γ describes a closed curve in an α-plane then this connection is integrable. It is clear that this connection preserves the conformal metric and it is straightforward to check that it actually preserves the full metric. Thus if the connection is torsion-free, it *is* the metric connection, is flat on α-planes and provides a global parallelism of primed spinors, whence the Ricci tensor vanishes and the Weyl tensor is anti-self-dual. This completes the proof that curved twistor spaces $\mathcal{T}$ (satisfying (i)–(iv)) lead to half-flat space-times $\mathbf{M}$, if we can establish that the connection defined by figure 12.3 is torsion-free. This is a small technical point and we remove it to an exercise.

The final question we face is that of constructing curved twistor spaces. Any method that generates manifolds satisfying (i)–(iv) will do (for example Hitchin 1979) but we shall consider the problem of starting with flat twistor space $\mathbf{T}$ and arriving at $\mathcal{T}$ by a deformation of the complex structure of $\mathbf{T}$ (more accurately, of *part* of $\mathbf{T}$). As usual, we shall be informal in our treatment of holomorphic deformations. A fuller account adapted to the problems of twistor theory may be found in Burns (1979).

The simplest way to define a deformation of a complex manifold X is in terms of a coordinate atlas $\{U_i\}$ with coordinate functions Z_i and transition relations

$$Z_i = f_{ij}(Z_j) \text{ on } U_i \cap U_j.$$

The manifold X_t is a holomorphic deformation of X if X_t has the same underlying real manifold and atlas but the transition relations are holomorphic also in t:

$$Z_i = f_{ij}(Z_j, t), \ f_{ij}(Z_j, 0) = f_{ij}(Z_j). \tag{12.9}$$

Given (12.9) the deformation is defined to first order by the partial derivative of the transition relations with respect to t. In fact this infinitesimal deformation is correctly described by a vector field cocycle. If X is n-dimensional and $\alpha = 1, \ldots, n$ is the coordinate index then (12.9) stands for

$$Z^\alpha_{\ i} = f^\alpha_{\ ij}(Z^\beta_{\ j}, t) \text{ on } U_i \cap U_j.$$

Define

$$\theta_{ij} = \left.\frac{\partial Z_i^\alpha}{\partial t}\right|_{t=0} \frac{\partial}{\partial Z_i^\alpha} = \frac{\partial f_{ij}^\alpha}{\partial t}\frac{\partial}{\partial Z_i^\alpha} \tag{12.10}$$

when there is an implied summation on the Greek index but not on the Latin index. (Also, for notational convenience, evaluation at $t = 0$ will be understood.)

θ_{ij} is evidently a 1-cochain of vector fields; in fact it is a 1-cocycle. To see this, note that on a triple overlap we have

$$\begin{aligned} Z_j^\alpha &= f_{jk}^\alpha(Z_k^\beta, t) \\ Z_i^\alpha &= f_{ij}^\alpha(Z_j^\beta, t) = f_{ij}^\alpha(f_{jk}^\beta(Z_k^\gamma, t), t) \end{aligned}$$

so that

$$\theta_{jk} = \frac{\partial f_{jk}^\alpha}{\partial t} \frac{\partial}{\partial Z_j^\alpha}$$

$$\theta_{ik} = \frac{\partial f_{ij}^\alpha}{\partial t} \frac{\partial}{\partial Z_i^\alpha} + \frac{\partial f_{ij}^\alpha}{\partial Z_j^\beta} \frac{\partial f_{jk}^\beta}{\partial t} \frac{\partial}{\partial Z_i^\alpha}$$

$$= \theta_{ij} + \theta_{jk}$$

since

$$\frac{\partial f_{ij}^\alpha}{\partial Z_j^\beta} \frac{\partial}{\partial Z_i^\alpha} = \frac{\partial}{\partial Z_j^\beta}$$

whence $\theta_{ij} + \theta_{jk} + \theta_{ki} = 0$. Now if each Z_i^α is subjected to a t-dependent coordinate transformation on U_i so that

$$Z_i^\alpha \to \hat{Z}_i^\alpha = h_i^\alpha(Z_i^\beta, t); \ h_i^\alpha(Z_i^\beta, 0) = Z_i^\alpha,$$

then

$$\hat{Z}_i^\alpha = h_i^\alpha(f_{ij}^\beta(h_j^{-1\,\gamma}(\hat{Z}_j^\delta, t), t), t)$$

so that

$$\hat{\theta}_{ij} = \frac{\partial \hat{Z}^\alpha}{\partial t} \frac{\partial}{\partial \hat{Z}_i^\alpha}$$

$$= \frac{\partial h_i^\alpha}{\partial t} \frac{\partial}{\partial Z_i^\alpha} + \frac{\partial h_i^\alpha}{\partial Z_i^\beta} \frac{\partial Z_i^\beta}{\partial t} \frac{\partial}{\partial Z_i^\alpha} - \frac{\partial h_i^\alpha}{\partial Z_i^\beta} \frac{\partial Z_i^\beta}{\partial Z_j^\gamma} \frac{\partial h_j^\gamma}{\partial t} \frac{\partial}{\partial Z_i^\alpha}$$

$$= \xi_i + \theta_{ij} - \xi_j$$

where

$$\xi_i = \frac{\partial h_i^\alpha}{\partial t} \frac{\partial}{\partial Z_i^\alpha}.$$

Here ξ defines a 0-cochain of vector fields. Thus the coordinate-independent description of an infinitesimal deformation is as a 1-cochain modulo coboundaries, i.e. as an element of $\check{H}^1(X; \Theta)$ where Θ is the sheaf of germs of holomorphic vector fields. (See figure 12.4.)

As an example, we consider the case $X = \mathbf{P}^1$. Since X is one-dimensional, Θ is a line bundle (see exercise 3g) and we have only to decide which one. A vector field Y can be written in homogeneous coordinates as $Y = \alpha_{A'}(\pi_{B'}) \frac{\partial}{\partial \pi_{A'}}$ where $\alpha_{A'}$ is a pair of functions homogeneous of degree 1. Furthermore the vector field $\pi_{A'} \frac{\partial}{\partial \pi_{A'}}$ on $\mathbf{C}^2$ represents the zero vector

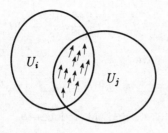

Figure 12.4. The vector field θ_{ij} defined on $U_i \cap U_j$.

field on $\mathbf{P}^1$ so we only care about the part of $\alpha_{A'}$ *not* proportional to $\pi_{A'}$, i.e. we must consider

$$\alpha_{[A'}\pi_{B']} = \frac{1}{2}\epsilon_{A'B'}\alpha_{C'}\pi^{C'} = \frac{1}{2}\epsilon_{A'B'}V(\pi).$$

Thus Y is determined by the single function V which is homogeneous of degree 2 in $\pi_{A'}$ and so $\Theta \cong \mathcal{O}(2)$. By the argument above, the infinitesimal deformations of $\mathbf{P}^1$ are in one-to-one correspondence with the group $\check{H}^1(\mathbf{P}^1; \Theta) = \check{H}^1(\mathbf{P}^1; \mathcal{O}(2))$. However, by the results of chapter 9, we know that this group is zero, so there are no infinitesimal deformations of $\mathbf{P}^1$.

In fact this result holds generally for $\mathbf{P}^n$ (Morrow and Kodaira 1971), so we must consider deformations of only *part* of twistor space, typically the neighbourhood of one particular line, L_p.

The deformations which we seek must preserve conditions (i)–(iv). To preserve the first, we must restrict to vector fields tangent to the fibres of π:

$$Y = f^A(\omega, \pi)\frac{\partial}{\partial \pi^A}. \qquad (12.11)$$

For (ii) we demand that Y and Υ commute:

$$[\Upsilon, \Upsilon] = (\Upsilon f^A - f^A)\frac{\partial}{\partial \omega^A} = 0$$

which means that f^A must be homogeneous of degree 1 in $(\omega^A, \pi_{A'})$.

If we think of Y as defined on the overlap between two sets $U, \hat{U}$ with coordinates $\omega^A, \hat{\omega}^A$, then the infinitesimal deformation is given by

$$\hat{\omega}^A = \omega^A + tf^A(\omega^B, \pi_{B'}) + O(t^2) \qquad (12.12)$$

so that, at fixed $\pi_{A'}$,

$$d\hat{\omega}^A = d\omega^A + t\frac{\partial f^A}{\partial \omega^B}d\omega^B + O(t^2). \qquad (12.13)$$

To satisfy (iii) we must have

$$\hat{\mu} = \epsilon_{AB}d\hat{\omega}^A \wedge d\hat{\omega}^B = \epsilon_{AB}d\omega^A \wedge d\omega^B = \mu.$$

Substituting from (12.13) we find this reduces to

$$\epsilon_A{}^B\frac{\partial f^A}{\partial \omega^B} = 0$$

which implies $f^A = \epsilon^{AB}\frac{\partial f}{\partial \omega^B}$ for some $f(\omega^C, \pi_{C'})$ homogeneous of degree 2.

Thus we satisfy conditions (i)–(iii) of five pages ago by taking a 1-cocycle $f \in \check{H}^1(X; \mathcal{O}(2))$ and forming the vector field cocycle

$$Y = \epsilon^{AB}\frac{\partial f}{\partial \omega^A}\frac{\partial}{\partial \omega^B}.$$

From chapter 8, we recall that such an f is precisely the 'twistor function' for a linearised gravitational field. Here we are seeing this cocycle in an active role, determining an infinitesimal deformation. To define a finite deformation, we consider the case when the cocycle is defined on a two-set cover of the neighbourhood of a line. If the sets are $U, \hat{U}$ with coordinates $\omega^A, \hat{\omega}^A$ we first seek the integral curves of Y, i.e. we solve the equations

$$\frac{d\omega^A}{dt} = -\epsilon^{AB}\frac{\partial f}{\partial \omega^B} \qquad (12.14)$$
$$\omega^A(0) = \omega^A$$

and then set

$$\hat{\omega}^A = \omega^A(t). \qquad (12.15)$$

Here t plays the rôle of the deformation parameter. (We should emphasize that this is *one* way of getting a finite deformation from an infinitesimal one, there are others, and further this method depends on the choice of representative of the cohomology class of f. Although the *infinitesimal* deformation is a cohomological object, the *finite* deformation isn't.)

So far we have neglected condition (iv). However, there are some theorems of Kodaira (Morrow and Kodaira 1971; Hansen et al. 1978) which assure us that we are justified in doing so, that in fact (iv) will be true automatically for small enough deformations (i.e. for the deformation parameter t in a neighbourhood of zero). The relevant theorems are as follows:

1. Suppose X_t is a holomorphic family of complex manifolds for t in a neighbourhood U_1 of zero in $\mathbf{C}$, Y is a complex submanifold of X_0 and N_0 is the normal bundle of Y in X_0. Then there exists a neighbourhood U_2 of zero in $\mathbf{C}$ and a holomorphic family of complex submanifolds $Y_t \subset X_t$ for $t \in U_2$ with $Y_0 = Y$ provided $\breve{H}^1(Y; N_0) = 0$.

2. A sufficient condition for Y_t to lie in an n-dimensional family of submanifolds in X_t is $\breve{H}^1(Y_t; N_t) = 0$ and then $n = \dim \breve{H}^0(Y_t; N_t)$.

We want to use these when Y_0 is a particular $\mathbf{P}^1$, say L_p, in $\mathbf{PT}$ and X_0 is a neighbourhood of Y_0. We know already that $N_0 \cong \mathcal{O}(1) \oplus \mathcal{O}(1)$ so that $H^1(Y_0; N_0) = 0$. Thus by Theorem 1 above, there is an analogue Y_t of Y_0 in X_t. If we knew that $N_t \cong N_0$ then, by Theorem 2, Y_t would lie in an n-dimensional family where

$$n = \dim \breve{H}^0(Y_t; N_t) = \dim \breve{H}^0(Y_0; N_0) = 4.$$

Here we have used the two facts that Y_t and Y_0 are the same as complex manifolds, namely $\mathbf{P}^1$, since there are no infinitesimal deformations of $\mathbf{P}^1$, and that 4 is the answer in $\mathbf{PT}$.

In this case, we would know not only that there was a four-parameter family, but also that the conformal structure was preserved.

Since there are no deformations of $\mathbf{P}^1$, to see what happens to N_t we must consider the question of deforming vector bundles over $\mathbf{P}^1$. If the transition matrices of a vector bundle E in some trivialisation are allowed to depend on a parameter t say, so that

$$g_\alpha{}^\beta = g_\alpha{}^\beta(Z, t)$$

neglecting the cochain indices) then the infinitesimal deformation is defined by

$$\frac{\partial}{\partial t} g_\alpha{}^\beta \bigg|_{t=0}$$

which is a 1-cocycle of endomorphisms of E. By a similar argument to the previous one, what is actually involved is a cohomology class in $H^1(\mathbf{P}^1; \mathrm{End}(E))$ where $\mathrm{End}(E)$ is the sheaf of germs of endomorphisms of E. When $E = \mathcal{O}(1) \oplus \mathcal{O}(1)$, an endomorphism is locally a 2×2 matrix of homogeneity zero functions so that

$$\breve{H}^1(\mathbf{P}^1; \mathrm{End}(E)) = \oplus_{4 \text{ copies}} \breve{H}^1(\mathbf{P}^1; \mathcal{O}) = 0.$$

Thus in this case there are no infinitesimal deformations of N_0 and so, for t in a neighbourhood of zero, $N_t \cong N_0 = \mathcal{O}(1) \oplus \mathcal{O}(1)$. (This neighbourhood of zero may be strictly smaller than the one supplied by Theorem 1.)

All we needed for Theorem 2 was

$$\check{H}^1(\mathbf{P}^1; N_t) = 0; \ \dim \check{H}^0(\mathbf{P}^1; N_t) = 4. \tag{12.16}$$

If $N_t = \mathcal{O}(1) \oplus \mathcal{O}(1)$, these are certainly satisfied and furthermore the quadratic conformal structure is still defined. However, it could happen that for some value of t the analogue Y_t of Y_0 persisted but N_t became $\mathcal{O}(2) \oplus \mathcal{O}$ or even $\mathcal{O}(3) \oplus \mathcal{O}(-1)$. Now (12.16) is still satisfied so that Y_t still lies in a four-dimensional family but the quadratic conformal structure is not defined. This would be reflected in the space-time $\mathbf{M}$ corresponding to X_t by a singularity in the conformal structure. These questions are considered further in Tod (1982).

Armed with these theorems of Kodaira, then, we may construct deformed twistor spaces by forming the transition relations (12.15) and then seeking the four-parameter family of holomorphic curves whose existence, at least for small t, is guaranteed.

We conclude this section with a simple example of this process due originally to G.A.J. Sparling. Sparling considers the 1-cocycle

$$f = \frac{(\omega^0)^4}{4\pi_{0'}\pi_{1'}}$$

defined with respect to the two-set cover of $\mathbf{PT}^+$ (Ward 1978):

$$U = \{Z^\alpha \in \mathbf{PT}^+ - I : \pi_{0'} \neq 0\}$$
$$\hat{U} = \{Z^\alpha \in \mathbf{PT}^+ - I : \pi_{1'} \neq 0\}.$$

The line I, where $\pi_{A'} = 0$, is removed. (12.14) becomes

$$\frac{d\omega^0}{dt} = 0; \ \frac{d\omega^1}{dt} = \frac{(\omega^0)^3}{\pi_{0'}\pi_{1'}}$$

whence

$$\hat{\omega}^0 = \omega^0; \ \hat{\omega}^1 = \omega^1 + t\frac{(\omega^0)^3}{\pi_{0'}\pi_{1'}}. \tag{12.17}$$

These are the transition relations. Note that ω^0 defines a global function, homogeneous of degree 1. On any holomorphic curve L_p, this function must be a linear polynomial in $\pi_{A'}$:

$$\omega^0 = \xi\pi_{0'} + u\pi_{1'}, \ u, \xi \in \mathbf{C}. \tag{12.18}$$

Substituting (12.18) in (12.17) and rearranging:

$$\hat{\omega}^1 - \frac{t}{\pi_{1'}}(\xi^3(\pi_{0'})^2 + 3\xi^2 u\pi_{0'}\pi_{1'} + 3\xi u^2(\pi_{1'})^2) = \omega^1 + tu^3\frac{(\pi_{1'})^2}{\pi_{0'}}. \tag{12.19}$$

This defines a global function homogeneous of degree 1, in other words another linear polynomial in $\pi_{A'}$, on L_p so

$$\omega^1 = -tu^3\frac{(\pi_{1'})^2}{\pi_{0'}} + v\pi_{0'} + \tilde{\xi}\pi_{1'}, \ v, \tilde{\xi} \in \mathbf{C} \tag{12.20}$$

and (12.18) with (12.20) give the holomorphic curves in terms of the four parameters $x^a = (u, v, \xi, \tilde{\xi})$.

If two nearby curves C_1, C_2 are given by x^a and $x^a + \delta x^a$ then the vector $v^a = \delta x^a$ is null if the equations

$$\delta\omega^0 \equiv \pi_{0'}\delta\xi + \pi_{1'}\delta u = 0$$
$$\delta\omega^1 \equiv \pi_{0'}\delta v + \pi_{1'}\delta\tilde{\xi} - 3t\frac{(\pi_{1'})^2}{\pi_{0'}}u^2\delta u = 0$$

have a simultaneous solution. The condition for this is

$$\delta u\delta v - \delta\xi\delta\tilde{\xi} - 3tu^2(\delta\xi)^2 = 0 \tag{12.21}$$

so that this is the conformal metric, and it is quadratic as anticipated.

For the full metric, we consider two null vectors V^a and U^a satisfying (12.21). For simplicity, we may take $V^a = (\delta u, 0, 0, 0)$; $U^a = (0, \Delta v, 0, 0)$. Correspondingly $\delta\omega^A$ and $\Delta\omega^A$ are given by

$$\delta\omega^0 = \pi_{1'}\delta u$$
$$\delta\omega^1 = -3t\frac{(\pi_{1'})^2}{pi_{0'}}u^2\delta u$$
$$\Delta\omega^0 = 0$$
$$\Delta\omega^1 = \pi_{0'}\Delta v.$$

By (12.8), the contraction $g_{ab}U^aV^b$ is given by

$$g_{ab}U^aV^b = \frac{\epsilon_{AB}\delta\omega^A\Delta\omega^B}{\pi_{0'}\pi_{1'}}$$
$$= \delta u\Delta v$$
$$= g_{01}\delta u\Delta v$$

so that $g_{01} = 1$ and finally the metric may be written

$$g_{ab}dx^a dx^b = 2du\,dv - 2d\xi\,d\tilde{\xi} - 6tu^2d\xi^2. \tag{12.22}$$

It may be checked (if desired!) that this rather simple metric does indeed satisfy Einstein's vacuum equations. The Weyl tensor is anti-self-dual and in fact Type N.

We have called this chapter the Non-Linear Graviton since this is the name which has become attached to this construction. Penrose's motivation for this name is his suggestion (Penrose 1976) that these spaces should form the one-particle states of a future quantum theory of gravity. Each state is genuinely a non-linear object, in the sense that it has some curvature, rather than a solution of a linear field equation in Minkowski space.

As he also emphasised at the time, however, he would wish to preserve the name non-linear graviton for those deformed twistor spaces which were actually positive-frequency in an appropriate sense. The appropriate sense would be related to the space T being a deformation of the whole top half $\mathbf{T}^+$ of twistor space. Each fibre would have to have a boundary corresponding to a deformed version of the line I and these would be related from fibre to fibre.

The implications of this extra restriction have yet to be fully worked out.

Exercises 12

a) Suppose E is a holomorphic vector bundle over a complex manifold X and s is a holomorphic section of E which vanishes to first order on a submanifold Y of X (i.e. $s = 0$, $ds \neq 0$ on Y). Show that the normal bundle of Y in X is canonically $E \mid_Y$. (Now if E is the bundle $\mathcal{O}(1) \oplus \mathcal{O}(1)$ over $\mathbf{CP}^3$, a section vanishes on a line, so that the normal bundle of a line in $\mathbf{CP}^3$ is $\mathcal{O}(1) \oplus \mathcal{O}(1)$. We are grateful to Dr. M.G. Eastwood for this exercise.)

b) For a two-set cover of the neighbourhood N of a line in $\mathbf{PT}$, the linearised version of (12.14) is

$$\hat{\omega}^A = \omega^A - \lambda \epsilon^{AB} \frac{\partial f}{\partial \omega^B}$$

with $f \in \check{H}^1(N; \mathcal{O}(2))$.

Find the linearised version of the four-parameter family of cross-sections and the metric and show that the linearised curvature is proportional to what would be obtained by treating f as a twistor function in the contour integrals of chapter 10.

c) Show that the connection defined geometrically in this section is torsion-free.

Chapter 13

Penrose's Quasi-Local Momentum and Angular Momentum

In general relativity there is a long-standing problem of defining momentum and angular momentum in a general curved space-time. In this chapter, we describe an application of twistor theory which aims to provide a solution to this problem. As we shall see, this aim is not entirely achieved and the programme has both successes and failures.

We recall that in general relativity, all the local matter content is described by the stress–energy tensor T_{ab}. Gravitational energy, whether in gravitational waves or in the form of gravitational potential energy, is notoriously non-local and one cannot expect to characterise it by a local density. Instead, Penrose (1982) has suggested that one should seek a non-local invariant associated to any two-surface S and representing the total momentum–angular momentum flux through that surface. This non-local invariant will be constructed by twistorial techniques tailored to give the right answer for linearised general relativity where there is a clear right answer.

We begin by reviewing the definition of momentum and angular momentum in special relativity and in linearised general relativity.

A material system in special relativity is defined by its stress–energy tensor T_{ab} which we may suppose for definiteness to have support within a world tube W in Minkowski space, $\mathbf{M}$. (The case of say electromagnetic fields spread throughout $\mathbf{M}$ is a simple generalisation.)

Conservation of energy is expressed by the condition that T_{ab} has zero divergence:

$$\nabla^a T_{ab} = 0. \qquad (13.1)$$

Given any Killing vector k^b the current $J_a = T_{ab}k^b$ is then conserved:

$$\nabla^a J_a = 0$$

and we may define a conserved charge by integration over an arbitrary 3-surface Σ:

$$Q[k] = \int_\Sigma T_{ab}k^b d\sigma^a. \qquad (13.2)$$

Here we are thinking of Q as a linear functional on the space of all Killing vectors. Q has ten real components corresponding to four components of momentum, when k^b is a translation, and six components of angular momentum, when k^b is a rotation.

In fact all the information of total momentum and angular momentum is contained in (13.2) when k^b is a self-dual (s.d.) Killing vector, that is, one for which the derivative $\nabla_a k_b$ is a self-dual two form. These comprise the four translations and the s.d. rotations, which are necessarily complex.

In linearised general relativity, the source T_{ab} produces a linearised Weyl tensor, which can be thought of as a spin-2 field on $\mathbf{M}$ related to T_{ab} by the linearised Bianchi identity

$$\nabla_{A'}{}^A \psi_{ABCD} = \nabla_{(B}{}^{B'} \Phi_{CD)A'B'} \qquad (13.3)$$

where $\Phi_{ab} = \frac{1}{2}(T_{ab} - \frac{1}{4}g_{ab}T^c{}_c)$. Now one anticipates being able to measure the momentum and angular momentum of the source with a two-surface integral involving the gravitational field in an analogous way to the formula in Newtonian theory for the total mass:

$$M = \int_\Sigma \rho dV = \frac{1}{4\pi G} \oint_S \nabla \varphi.d\underline{s}$$

The two expressions, the 2-surface integral and the 3-surface integral, are related by the divergence theorem.

Here the desired formula is

$$Q[k] = \frac{i}{4\pi G} \oint_S R_{abcd} f^{cd} d\sigma^{ab} = \int_\Sigma T_{ab}k^b d\sigma^a \qquad (13.4)$$

where R_{abcd} is the linearised Riemann tensor, f^{ab} is a bivector potential for the Killing vector k^b, in a sense to be explained, and $d\sigma^{ab}$ is the volume form for S.

We saw in exercise 4e that if ω^{AB} is a solution of the twistor equation

$$\nabla_{A'}{}^{(A}\omega^{BC)} = 0 \tag{13.5}$$

then

$$\nabla_{A'A}\omega_{BC} = -i\epsilon_{A(B}k_{C)A'} \tag{13.6}$$

where k_b turns out to be a s.d. Killing vector.

In coordinates the solution of(13.5) is

$$\omega^{AB}(x) = -i\varphi^{A'B'}x^A{}_{A'}x^B{}_{B'} + iT^{A'(A}x^{B)}{}_{A'} + \Omega^{AB} \tag{13.7}$$

and from (13.6)

$$k^{AA'} = T^{AA'} - 2i\varphi^{A'}{}_{B'}x^{B'A}. \tag{13.8}$$

Thus to make (13.4) work, one takes

$$f^{ab} = \omega^{AB}\epsilon^{A'B'} \tag{13.9}$$

where ω^{AB} is related to k^b by (13.6). It is then a simple exercise to verify (13.4). Another way of looking at (13.4) is in terms of spin-lowering (exercise 4d). Outside the world tube W, ψ_{ABCD} is a z.r.m. field. For any symmetric twistor $T^{\alpha\beta}$ corresponding to a field ω^{CD}, it follows that

$$\varphi_{AB} = \psi_{ABCD}\omega^{CD}$$

is a solution of Maxwell's equations. We may therefore calculate its charge by integration over any two-surface S. This gives a charge for each symmetric twistor $T^{\alpha\beta}$ and so defines a twistor $A_{\alpha\beta}$. Then (13.4) relates $A_{\alpha\beta}$, the *momentum–angular momentum twistor* or kinematic twistor of the source, to explicit integrals over the source.

Not every twistor $A_{\alpha\beta}$ can arise as a kinematic twistor (Penrose and MacCallum 1973). Note that if the chosen element ω^{AB} of $T \otimes_S T$ is actually a constant spinor field on **M**, then the corresponding charge vanishes. Further if ω^{AB} is such that k_a is a real translation, the corresponding charge is real.

To see what this means for $A_{\alpha\beta}$ we substitute (13.7) and (13.8) into (13.4). If the matrix of $A_{\alpha\beta}$ is

$$A_{\alpha\beta} = \begin{bmatrix} \lambda_{AB} & P_A{}^{B'} \\ P_B{}^{A'} & -2i\mu^{A'B'} \end{bmatrix}$$

then

$$
\begin{aligned}
Q[k] &= A_{\alpha\beta}T^{\alpha\beta} \\
&= \lambda_{AB}\Omega^{AB} + P_A{}^{A'}T^A{}_{A'} + 2i\mu^{A'B'}\varphi_{A'B'} \\
&= \int T_{ab}(T^b + 2i\varphi^{B'C'}x^B{}_{C'})d\sigma^a
\end{aligned}
$$

so that $\lambda_{AB} = 0$, $P_a = \overline{P}_a$ is the momentum and $\mu^{A'B'}$ is the s.d. angular momentum. These restrictions on $A_{\alpha\beta}$ can be written concisely with the aid of the infinity twistor $I^{\alpha\beta}$ as

$$A_{\alpha\beta}I^{\beta\gamma} = \overline{A}^{\gamma\delta}I_{\delta\alpha}. \qquad (13.10)$$

It is these conditions which reduce the ten complex components of $A_{\alpha\beta}$ to the four real components of momentum and three complex components of angular momentum. Note that one also has the formula

$$A_{\alpha\beta}\overline{A}^{\alpha\beta} = -2P_aP^a = -2m^2 \qquad (13.11)$$

for the total mass threading through S. The formula (13.11) for the mass requires for its definition the norm on twistor space, while the formula (13.10), which reduces the number of components to ten real ones and distinguishes the components of $A_{\alpha\beta}$ which are the momentum, requires in addition the infinity twistor. These points will be significant later on.

In a general curved space, there are no Killing vectors and a formula like (13.2) is of no use. However, one might be able to use the first equality in (13.4) if one had a way of recognising or defining f^{ab} solely at S. We consider this problem next.

Any solution of (13.5) is a linear combination of products of solutions to the valence one twistor equation

$$\nabla_{A'}{}^{(A}\omega^{B)} = 0; \quad \nabla_{A'A}\omega^B = -i\delta_A{}^B\pi_{A'}. \qquad (13.12)$$

An arbitrary 2-surface S defines a pair of spinors (o_A, ι_A) tangent respectively to the out- and in-going null directions orthogonal to S. If we normalise the dyad by $o_A\iota^A = 1$ then there remains the freedom

$$(o_A, \iota_A) \to (\lambda o_A, \lambda^{-1}\iota_A)\ \lambda \in \mathbf{C}^*. \qquad (13.13)$$

The complex null vector $m^a = o^A\bar{\iota}^{A'}$ is tangent to S and we may consider the parts of (13.12) which involve derivatives only tangential to S. These are

$$o_B\delta\omega^B = 0 \qquad ; \quad \iota_B\overline{\delta}\omega^B = 0; \qquad (13.14)$$

$$o_B\overline{\delta}\omega^B = -i\overline{o}^{A'}\pi_{A'} \quad ; \quad \iota_B\delta\omega^B = i\overline{\iota}^{A'}\pi_{A'}, \qquad (13.15)$$

where $\delta = m^a\nabla_a$.

Here (13.14) is to be viewed as a pair of equations on the two components of ω^A, and (13.15) as the definition of $\pi_{A'}$, in terms of ω^A. We shall refer to (13.14), (13.15) as *the two-surface twistor equations* and to a solution $(\omega^A, \pi_{A'})$ as a *two-surface twistor*. Note that (13.14) defines an elliptic system for ω^A. (This is apparent when the equations are written out in coordinates. (Essentially it is because δ is closely related to the $\bar{\partial}$-operator of the complex structure of S, which is elliptic; see exercise 9d and Wells (1980).)

Before embarking on a more detailed disussion of (13.14) and (13.15) we describe a formalism introduced by Geroch, Held and Penrose (1973: the GHP formalism) precisely adapted to the situation considered here, namely where a spinor dyad is specified up to the freedom (13.13).

We suppose that the dyad is chosen in a four-dimensional neighbourhood of S. As a first step all spinorial and vectorial quantities are replaced by the set of scalars which are their components in the chosen dyad. These scalars are weighted, in that they change under transformation (13.13) of the dyad and, specifically, a scalar η is said to have weights (p, q) if it transforms as

$$\eta \to \hat{\eta} = \lambda^p \bar{\lambda}^q \eta. \tag{13.16}$$

Next, separate letters are introduced for the components of the spinor connection which are weighted scalars. These are

$$
\begin{array}{llll}
\kappa & = & o^A D o_A; & \kappa' & = & -\iota^A \Delta \iota_A \\
\rho & = & o^A \bar{\delta} o_A; & \rho' & = & -\iota^A \delta \iota_A \\
\sigma & = & o^A \delta o_A; & \sigma' & = & -\iota^A \bar{\delta} \iota_A \\
\tau & = & o^A \Delta o_A; & \tau' & = & -\iota^A D \iota_A
\end{array}
\tag{13.17}
$$

where $D = l^a \nabla_a$, $\Delta = n^a \nabla_a$, $\delta = m^a \nabla_a$. It can be checked that these are weighted scalars, whose weights are left as an exercise. (Some of the spin coefficients (13.17) have already been encountered in chapter 6.)

The prime introduced in (13.17) indicates a symmetry generated by the transformation

$$o^A \to i\iota^A; \quad \iota^A \to io^A; \quad \bar{o}^{A'} \to -i\bar{\iota}^{A'}; \quad \bar{\iota}^{A'} \to -i\bar{o}^{A'} \tag{13.18}$$

under which e.g.

$$\rho \to \rho'; \quad \rho' \to \rho$$

and so on. Note in particular that $\delta' = \bar{\delta}$. This symmetry is systematically exploited in the formalism and effectively halves the number of equations which need to be written out. The remaining parts of the spin connection are

$$
\begin{array}{llll}
\beta & = & \iota^A \delta o_A; & \beta' & = & -o^A \bar{\delta} \iota_A \\
\epsilon & = & \iota^A D o_A; & \epsilon' & = & -o^A \Delta \iota_A.
\end{array}
\tag{13.19}
$$

These do not transform as weighted scalars but can be combined with the operators D, Δ, δ and $\bar{\delta}$ to produce weighted operators $\wp$ and $\eth$ pronounced, 'thorn' and 'eth' respectively, according to the scheme

$$
\begin{array}{llll}
\text{\th}\eta &=& (D - p\epsilon - q\bar{\epsilon})\eta; & \text{\th}'\eta &=& (\Delta + p\epsilon' + q\bar{\epsilon}')\eta \\
\eth\eta &=& (\delta - p\beta + q\bar{\beta}')\eta; & \eth'\eta &=& (\delta' + p\beta' - q\bar{\beta})\eta
\end{array}
\tag{13.20}
$$

when acting on a quantity η of weight (p, q).

Again it is left as an exercise to show that these operators have well-defined weights and to calculate what they are. (To say that e.g. $\text{\th}$ has weight $(1, 1)$ is to say that $\text{\th}\eta$ has weight $(p+1, q+1)$ if η has weight (p, q).)

To write (13.14) and (13.15) in the GHP formalism we introduce the scalars ω^0 and ω^1 of weight $(-1, 0)$ and $(1, 0)$ respectively which are the components of ω^A, and the scalars $\pi_{0'}$ and $\pi_{1'}$ of weights $(0, 1)$ and $(0, -1)$ which are the components of $\pi_{A'}$:

$$
\omega^A = \omega^0 o^A + \omega^1 \iota^A; \quad \pi_{A'} = \pi_{1'}\bar{o}_{A'} - \pi_{0'}\bar{\iota}_{A'}.
\tag{13.21}
$$

Then (13.14) and (13.15), the two-surface twistor equations, become

$$
\eth\omega^1 - \sigma\omega^0 = 0; \qquad \eth'\omega^0 - \sigma'\omega^1 = 0
\tag{13.22}
$$

$$
\eth'\omega^1 - \rho\omega^0 = -i\pi_{0'}; \qquad \eth\omega^0 - \rho'\omega^1 = -i\pi_{1'}.
\tag{13.23}
$$

In each case, the second equation is the prime of the first and of course every term in any one equation must have the same weight. These two observations provide useful checks to calculation!

The remainder of the GHP formalism consists of three sets of equations, which we relegate to an appendix. Firstly, there are six equations and their primes defining the curvature in terms of the spin coefficients. Next, and related to these, there are the commutators of the weighted operators. Finally there are the Bianchi identities.

As an exercise in the use of the first two sets of equations, we invite the reader to show that the commutator of $\eth$ and $\eth'$ applied successively to ω^0 and ω^1 in equations (13.22),(13.23) leads to the pair of new equations:

$$
\begin{array}{l}
\eth\pi_{0'} + \rho\pi_{1'} = i(\psi_2 - \varphi_{11} - \Lambda)\omega^1 + i(\psi_1 - \varphi_{01})\omega^0 \\
\eth'\pi_{1'} + \rho'\pi_{0'} = i(\psi_3 - \varphi_{21})\omega^1 + i(\psi_2 - \varphi_{11} - \Lambda)\omega^0.
\end{array}
\tag{13.24}
$$

These will be useful later. Returning to (13.22), we have claimed that this system represents an elliptic system from pairs of functions of weight $(-1, 0)$ and $(1, 0)$ to (by inspection) pairs of functions of weight $(-2, 1)$ and $(2, -1)$. We regard this as a map

$$
T : E \to F
$$

of vector bundles over S (see exercise 13b). Picking a Hermitian inner product on F, we may define the adjoint

$$T^\dagger : F \to E$$

and the index

$$\text{ind } T = \dim \ker T - \dim \ker T^\dagger.$$

By the Atiyah–Singer Index Theorem (see e.g. Palais 1965), this number is a topological invariant of the surface S and the bundles E, F; i.e. it is independent of the metric, connection and curvature of the space-time. In fact it turns out to be $4(1 - g)$ where g is the genus of S (Baston 1984). We shall generally be concerned only with topologically spherical surfaces, where the index is four. This implies that the solutions of (13.22) form a complex vector space $T(S)$ with dimension at least four. With more care, it can be shown that when S is 'near' to a metric sphere in flat space, $\dim T(S)$ is precisely four (Eastwood 1983) though examples are known where the dimension is higher (Jeffryes 1986). Thus we think of $T(S)$ as a twistor space defined at S by the two-surface twistor equations.

Returning to (13.4), this means that we may recognise the different choices of f_{ab} solely by solving (13.22) at S. The remaining components of the twistor equation (13.12) are then propagation equations for taking the two-surface twistors off the surface S. In a general curved space there will be curvature obstructions to doing this (since a space-time must be conformally flat to admit four linearly independent solutions to (13.12)). However, Penrose's proposal is to use (13.4) to define a kinematic twistor A at S even in curved space, by using for f_{ab} spinor fields constructed from $T(S)$. This defines A as a special element of the dual of $T(S) \otimes_S T(S)$. Writing out (13.4) in terms of curvature spinors in the GHP formalism we find:

$$A_{\alpha\beta} Z^\alpha_1 Z^\beta_2 = \frac{i}{4\pi G} \int \{ (\varphi_{01} - \psi_1)\omega^0_1 \omega^0_2 + (\varphi_{11} + \Lambda - \psi_2)(\omega^0_1 \omega^1_2 + \omega^1_1 \omega^0_2)$$
$$+ (\varphi_{21} - \psi_3)\omega^1_1 \omega^1_2 \} dS \qquad (13.25)$$

where ω^A_1, ω^A_2 are the spinor fields corresponding to the elements Z^α_1, Z^α_2 of $T(S)$. Here we are assuming that $T(S)$ is four-dimensional and introducing an index α for its elements. With the aid of (13.24) we find a simpler expression for (13.25):

$$A_{\alpha\beta} Z^\alpha_1 Z^\beta_2 = \frac{i}{4\pi G} \int \{ -\omega^0_1 (\eth \pi_{0'} + \rho \pi_{1'}) - \omega^1_1 (\eth' \omega_{1'} + \rho' \omega_{0'}) \} dS$$

$$= \frac{-i}{4\pi G} \int (\pi_{0'} \pi_{1'} + \pi_{0'} \pi_{1'}) dS \qquad (13.26)$$

where the second equality follows on integrating by parts.

At this stage, we have no norm on $T(S)$ and no infinity twistor so that we cannot ask whether (13.10) is satisfied, nor can we use (13.11) to calculate the total flux of energy through S. Put another way, the 'symmetry group' of $T(S)$ is still $GL(4, \mathbf{C})$ since we lack the norm to reduce it to $SU(2, 2)$ and the infinity twistor to reduce it further to the Poincaré group.

In an attempt to rectify this we consider the norm

$$\Sigma = \omega^A \overline{\pi}_A + \overline{\omega}^{A'} \pi_{A'} \tag{13.27}$$

appropriate to surfaces S in flat space.

The point is that, for such surfaces, Σ is a constant although constructed from fields ω^A and $\pi_{A'}$ which satisfy (13.22) and (13.23) and which therefore vary from point to point on S. One cannot prove the constancy of Σ solely from (13.22) and (13.23) however and this is a sticking point for surfaces in curved space.

We may proceed from here in two different ways: we may attempt to modify (13.27) so as to find a definition which is constant on an arbitrary S in curved space, or we may seek surfaces on which (13.27) is constant and content ourselves with a study of them. Following a suggestion of Penrose (1984), we call a surface S on which (13.27) is constant for every choice of two-surface twistor *non-contorted* so that if Σ varies, S is *contorted*. As we shall see below, a contorted surface is one which 'knows' that it is in the presence of conformal curvature. Our strategy will be first to say what is known in the non-contorted case, before a brief consideration of the contorted case.

Suppose then that the two-surface twistor space $T(S)$ has dimension four and that Σ defined by (13.27) is constant on S and so defines a norm on $T(S)$. What is special about such an S? First we observe that any point p of S defines a two-dimensional subspace L_p of $T(S)$ consisting of those fields ω^A which vanish at p. Since the space of two-planes in a four-dimensional complex vector space ($G(2, 4)$ in the terminology of exercise 3e) is known to be complexified compactified Minkowski space, $\mathbf{CM}^c$, this provides a map of S into $\mathbf{CM}^c$. This much is true regardless of Σ, but with the aid of Σ we distinguish $N(S)$ in $T(S)$, where Σ is zero, and hence real compactified Minkowski space.

Further the map found above maps S into this real Minkowski space. Since the construction is conformally invariant this suggests that if Σ is constant on S then S can be embedded in conformally flat space with the same first and second fundamental forms (since these are what enter into the two-surface twistor equation). The converse of this is evidently true, since if S can be so embedded, the two-surface twistors on S are the restrictions of twistors in the ambient space. We are thus lead to the proposition that Σ is constant on S iff S can be embedded in conformally flat space, and indeed this can be proved (Jeffryes 1984; Tod 1986). (Essentially, the constancy of Σ provides the integrability conditions for the conformal factor Ω and its derivatives $\flat\Omega$ and $\flat'\Omega$ at S.)

This then is what is special about non-contorted surfaces S, and for such surfaces we have a good definition of norm which we can use to calculate a total mass using (13.11). To verify (13.10) we need further to locate infinity in the real compactified Minkowski space associated to $T(S)$. To see that this is a separate problem over and above that connected with the norm consider a surface S in de Sitter space (exercises 5e and 7h). Here $T(S)$ is obtained by restricting twistors in the ambient space. The natural infinity twistor might be expected to be the non-simple one associated to de Sitter space, but the surface S cannot be expected to 'know' that it is not in Minkowski space!

We shall therefore begin with calculating the total mass for examples of non-contorted surfaces S.

The obvious first case to try is a sphere of symmetry in a spherically symmetric space-time (Tod 1983a). On such an S, we may choose the scaling of the dyad so that all GHP quantities are spherically symmetric, i.e. are constant. It follows from exercise 13b that all the ones with $p - q$ non-zero must vanish, leaving only ρ, ρ' and $N = \varphi_{11} + \Lambda - \psi_2$ as possibly non-zero real constants. Further these are dependent in that $K = 2(N - \rho\rho')$ is the Gauss curvature of S and so is related to the area Δ by the Gauss–Bonnet theorem, $K\Delta = 4\pi$. There is therefore essentially only one number distinguishing spheres of symmetry of the same area in different spherically symmetric space-times. Any such sphere is therefore equivalent to one of the same area in a de Sitter space with an appropriately chosen cosmological constant. This defines an embedding of S in a conformally flat space-time and so the norm is constant. To proceed further, we introduce a tetrad and coordinate system at S and write out the two-surface twistor equations explicitly.

The general spherically symmetric metric can be written

$$ds^2 = A^2(t,r)dt^2 - B^2(t,r)dr^2 - r^2(d\theta^2 + \sin^2 d\varphi^2) \qquad (13.28)$$

where r is an area distance. We define a null tetrad by

$$l^a\nabla_a = \frac{1}{\sqrt{2}}[A^{-1}\frac{\partial}{\partial t} - B^{-1}\frac{\partial}{\partial r}];$$

$$n^a\nabla_a = \frac{1}{\sqrt{2}}[A^{-1}\frac{\partial}{\partial t} + B^{-1}\frac{\partial}{\partial r}];$$

$$m^a \nabla_a \;=\; \frac{(1 + \zeta\overline{\zeta})}{r\sqrt{2}} \frac{\partial}{\partial \zeta}$$

where $\zeta = \tan\frac{\theta}{2} e^{i\varphi}$. Then l^a and n^a are orthogonal to a sphere S of constant r and t, and m^a is tangent to S. In this tetrad we find, as anticipated, that σ and σ' vanish while

$$\beta = -\overline{\beta}' = \frac{\overline{\zeta}}{2r\sqrt{2}}.$$

For (13.22) we obtain

$$\frac{\partial}{\partial \zeta}((1 + \zeta\overline{\zeta})^{\frac{1}{2}} \omega^1) \;=\; 0$$

$$\frac{\partial}{\partial \overline{\zeta}}((1 + \zeta\overline{\zeta})^{\frac{1}{2}} \omega^0) \;=\; 0$$

whence

$$\omega^1 = \frac{a + b\overline{\zeta}}{(1 + \zeta\overline{\zeta})^{\frac{1}{2}}}; \quad \omega^0 = \frac{c + d\zeta}{(1 + \zeta\overline{\zeta})^{\frac{1}{2}}} \tag{13.29}$$

and (a, b, c, d) are the four coordinates on $T(S)$, the two-surface twistor space of S.

Substituting (13.29) into (13.27) we find

$$\Sigma = \omega^A \overline{\pi}_A + \overline{\omega}^{A'} \pi_{A'} = \frac{-i}{r\sqrt{2}} (c\overline{b} - b\overline{c} + a\overline{d} - \overline{a}d). \tag{13.30}$$

As anticipated, the angular dependence has dropped out leaving a constant which is the expression for the norm in the coordinates (a, b, c, d).

Finally, we substitute in (13.20) to find the kinematic twistor:

$$\begin{aligned} A_{\alpha\beta} Z^\alpha Z^\beta \;&=\; \frac{i}{2\pi G} (\varphi_{11} + \Lambda - \psi_2) \int \omega^0 \omega^1 dS \\ &=\; \frac{i}{G} (\varphi_{11} + \Lambda - \psi_2)(ac + bd) r^2. \end{aligned} \tag{13.31}$$

Although it will have no Lorentz invariant significance, we may relabel the coordinates on $T(S)$ as

$$a = \Omega^0; \; b = \Omega^1; \; c = ir\sqrt{2}P_{1'}; \; d = -ir\sqrt{2}P_{0'}$$

so that (13.30) becomes

$$\Sigma = \Omega^0 \overline{P}_0 + \Omega^1 \overline{P}_1 + \overline{\Omega}^{0'} P_{0'} + \overline{\Omega}^{1'} P_{1'}$$

which is the norm in its usual form. Now (13.31) appears as

$$A_{\alpha\beta}Z^\alpha Z^\beta = 2P_A{}^{A'}\Omega^A P_{A'}$$

where

$$P_0^{1'} = -\frac{r^3 N}{G\sqrt{2}} = -P_1^{0'};\ P_0^{0'} = P_1^{1'} = 0$$

so

$$P^{00'} = P^{11'} = \frac{r^3 N}{G\sqrt{2}}.$$

Therefore the quasi-local mass defined by (13.11) is

$$M_P^2 = -\frac{1}{2}A_{\alpha\beta}\overline{A}^{\alpha\beta} = P_a P^a = (\frac{r^3 N}{G})^2,$$

i.e.

$$M_P = \frac{r^3 N}{G} = \frac{1}{G}(\frac{\Delta}{4\pi})^{\frac{3}{2}} N. \tag{13.32}$$

We have been led to a simple invariant expression (13.32) for the mass inside a surface S of area Δ in terms of Δ and the one remaining curvature component at S.

For the Schwarzschild space-time,

$$N = -\psi_2 = \frac{GM_S}{R^3}$$

where M_S is the Schwarzschild mass parameter. In this case (13.32) becomes

$$M_P = M_S$$

so that the Penrose mass at any sphere of symmetry in the Schwarzschild solution is just M_S, the Schwarzschild mass parameter. For the Reissner–Nordström space-time one calculates

$$N = \varphi_{11} - \psi_2 = G(\frac{M_S}{r^3} - \frac{e^2}{2r^4})$$

where e is the charge of the solution, and so

$$M_P = M_S - \frac{e^2}{2r}.$$

Here the mass has a contribution, due to the electromagnetic field, which depends on r. For a Reissner–Nördstrom black hole, the outer horizon is at

$$r = r_+ = M_S + (M_S^2 - e^2)^{\frac{1}{2}}.$$

At this radius

$$M_P = \frac{1}{2}r_+$$

which is the irreducible mass, i.e. the mass of a Schwarzschild black hole with the same area.

Finally, we consider the Friedmann–Robertson–Walker (FRW) cosmologies. These are spatially homogeneous and isotropic cosmological models with perfect fluid matter tensor. They have spatial cross-sections of constant curvature labelled $k = 1$ for the 3-sphere, $k = 0$ for flat space and $k = -1$ for hyperbolic space.

Now for M_p we find the product of the matter density with a volume, the volume of a sphere in flat space having the same surface area as the sphere S. For $k = 0$ this is straightforwardly density times volume within S but for the others it is different. In particular, one sees that in the $k = 1$ case the mass within a sphere of radius r increases from zero, attaining a maximum when the sphere is equatorial in the 3-sphere, and then decreases to zero again.

One may continue in this fashion and calculate M_p for various non-contorted two-surfaces in various space-times (Tod 1983a, 1990a). The results are broadly in agreement with physical intuition in a wide variety of circumstances. In particular, the original quasi-local mass definition successfully 'detects' the following aspects of mass and energy:

1. total mass-energy: in the sense of both the ADM and Bondi masses 'at infinity',

2. gravitational-wave-energy at infinity: as measured by the Bondi mass,

3. rest-mass energy: in, for example, the $k = 0$ FRW cosmology,

4. matter kinetic-energy: in the tilted Bianchi-type-V cosmology,

5. gravitational potential energy: in, for example, the time-symmetric initial-value-problem,

6. electro-static field energy: in the Reissner–Nordström solution,

7. gravitational-wave-energy 'quasi-locally': in cylindrical gravitational waves, and

8. the conserved quantity identified in the 'post-Newtonian' approximations to general relativity as energy.

(See Tod (1990a) for a review of the work which leads to this list, and for a bibliography of Penrose's quasi-local mass up to that date.) This is a remarkable list for a single, uniform definition to encompass. It is particularly striking that the definition, which was designed to work in that limit of general relativity which is linear theory ($G \to 0$), should also be correct in the other limit which is post-Newtonian theory ($c \to \infty$).

This completes the discussion of the non-contorted case and we turn to a brief discussion of contorted two-surfaces. To provide an example of these, we review the 'small-sphere' calculation (Kelly, Tod and Woodhouse 1986; Woodhouse 1987). The idea here is to calculate the quasi-local mass as a power series in the size of a sequence of small spheres, or small ellipsoids, as the surfaces shrink to a point.

The simplest way to construct such a sequence is as follows: first choose a point p in space-time M, and a unit time-like vector t^a at p; let the geodesics ruling the null cone N of p be generated by null vectors l^a which are normalised at the vertex p by

$$g_{ab}t^a l^b = 1.$$

If r is now an affine parameter determined by the parallelly-propagated generators l^a according to

$$l^a \nabla_a r = 1, \; r = 0 \text{ at } p$$

then the surfaces of constant t are 'small spheres'. We write $S(r)$ for the sphere of constant r. It is a straightforward, if lengthy, exercise to calculate the metric, connection and curvature of the space-time M on the null cone N as a power series in r. One may then solve (13.22) and (13.23) for the two-surface twistors, and calculate the kinematic twistor $A_{\alpha\beta}$ from (13.25) or (13.26) to any desired order in r.

In the presence of matter, the leading non-zero term in the kinematic twistor $A_{\alpha\beta}$ is $O(r^3)$. The norm defined by (13.27) is constant to $O(1)$, and the norm of $A_{\alpha\beta}$ to this order is the Minkowski-metric norm of the vector

$$P_a = \frac{4}{3}\pi r^3 T_{ab} t^b.$$

In other words, the quasi-local mass at this order is the norm of a momentum vector defined as the volume of the small sphere times the momentum-density obtained from the stress–energy tensor. This is an entirely satisfactory result, but is probably just a reflection of the fact that the quasi-local mass definition is correct for linear theory.

In a vacuum space-time, the leading non-zero term in $A_{\alpha\beta}$ is two orders higher, at $O(r^5)$. (Note therefore that if one divides by the volume of $S(r)$ the result vanishes at $r \to 0$; the *quasi-local* mass is *not* the integral of a density, which would be a *local* mass.) This time however the result of the calculation is quite unsatisfactory. The norm of the kinematic twistor can be negative even for small-spheres in the Schwarzschild solution; other invariants of $A_{\alpha\beta}$, such as its determinant, can be imaginary e.g. for small-spheres in the Kerr solution. At the same time, the norm defined by (13.27), although constant on $S(r)$ at $O(1)$ and $O(r)$, is varying at $O(r^2)$; the small-spheres are contorted at this order.

Before proceeding further, we make another observation about the contorted/non-contorted dichotomy. Given four linearly independent solutions to the two-surface equations, $(\omega^A{}_i, \pi_A{}^i)$ for $i = 1, 2, 3, 4$, we may define a determinant

$$\eta = \begin{vmatrix} \omega^0{}_1 & \omega^1{}_1 & \pi_{0'}{}^1 & \pi_{1'}{}^1 \\ \vdots & & & \vdots \\ \omega^0{}_4 & \omega^1{}_4 & \pi_{0'}{}^4 & \pi_{1'}{}^4 \end{vmatrix}. \tag{13.33}$$

For a two-surface S in flat or conformally flat space, and therefore also for a non-contorted S, this determinant η is constant. However, just as for the norm Σ, η cannot be shown to be constant just from the two-surface twistor equations (13.22, 13.23) and it will generally vary on a contorted S. Note that η has something of the character of a Jacobian. Penrose (1984; see also Penrose and Rindler 1984) has given arguments why the original definition of quasi-local quantities (13.4) should be modified by the inclusion of a multiple of η in the integrand.

Returning to the small-sphere calculation, we find that η from (13.33) is constant on $S(r)$ at $O(1)$ and $O(r)$ but varies at $O(r^2)$. Remarkably, if we modify the integral by including η as a factor in the integrand, as suggested by Penrose, then the integral vanishes at $O(r^5)$; in other words, this modification removes this pathology. The difficulty that remains is that there is still a leading non-zero term but it is now $O(r^6)$, and it has so far defied calculation.

We may summarise the situation as follows:

- There is a complete definition of quasi-local mass for non-contorted two-surfaces; it gives intuitively satisfying results in a wide variety of cases; however there are at present no general theorems about, for example, positivity.

- For contorted surfaces, there are indications that the definition is in need of modification, and suggestions for what the modifications should be; however there is not at this point a clearly satisfactory way forward.

Given the need of the original definition for modification, and given the attractiveness of the idea of a quasi-local mass definition, it is natural that other definitions have been formulated and investigated (see e.g. Ludvigsen and Vickers 1983; Bartnik 1989). To end this chapter, we shall consider one of these definitions which is quite close to the spirit of the original. This is the definition of Dougan and Mason (Dougan 1991; Dougan and Mason 1991) which is related to the definition of quasi-local charges for Yang–Mills theory (Tod 1983b) which was in turn inspired by Penrose's original definition of quasi-local quantities (Penrose 1982).

Given a space-like, topologically-spherical two-surface S and a spinor field λ_A on S, one considers the integral over S of the Witten–Nester integrand:

$$Q = \frac{i}{4\pi G} \int_S \lambda_{(A'} \nabla_{B')A} \overline{\lambda}^A - \overline{\lambda}^A \nabla_{A(A'} \lambda_{B')} d\sigma^{A'B'}. \tag{13.34}$$

This integrand is the 2-form which enters into the Witten proof of positive-energy (Witten 1981) and its subsequent extensions (e.g. Nester 1981; Horowitz and Perry 1982; Ludvigsen and Vickers 1982; Reula and Tod 1984; see also Penrose and Rindler 1984). As an exercise, the reader might show that the integral is real (integration by parts turns it into its complex conjugate).

For an asymptotically-constant spinor field $\lambda_{A'}$ and a two-surface S 'at infinity', (13.34) yields a component of the Bondi or ADM momentum. If one formally replaces $\overline{\lambda}^A$ by ω^A and $\lambda_{A'}$ by $\pi_{A'}$ where these are the components of a two-surface twistor then (13.34) becomes (13.26), though of course one has now given up the reality of (13.34). The idea of Dougan and Mason is to use (13.34) with spinor fields $\lambda^{A'}$ which are either 'holomorphic' or 'anti-holomorphic' on S in the sense that one or other of the equations

$$\overline{\delta}\lambda^{A'} = 0 \tag{13.35}$$

$$\delta\lambda^{A'} = 0 \tag{13.36}$$

which in the GHP formalism are

$$\overline{\eth}\lambda^{0'} - \rho'\lambda^{1'} = 0; \qquad \overline{\eth}\lambda^{1'} - \overline{\sigma}\lambda^{0'} = 0$$
$$\overline{\eth}\lambda^{0'} - \overline{\sigma}'\lambda^{1'} = 0; \qquad \eth\lambda^{1'} - \rho\lambda^{0'} = 0$$

holds on S. Each of these equations has a 2-dimensional complex vector space of solutions on S. Call these vector spaces $V, \tilde{V}$ respectively. We may label a basis of either by

$$\lambda^{A'}{}_{A'} = (\lambda^{A'}{}_{0'}, \lambda^{A'}{}_{1'})$$

and then the quantity

$$\epsilon_{0'1'} = \epsilon_{A'B'}\lambda_{0'}^{A'}\lambda_{1'}^{B'} \tag{13.37}$$

is easily seen to be constant on S (exercise). Provided it is non-zero, this therefore defines an $\epsilon_{A'B'}$ on each of V and $\tilde{V}$.

In more sophisticated language, equations (13.35, 13.36) define two distinct holomorphic structures on the spin-bundle $S^{A'}$ of the space-time at S; since $S^{A'}$ will generally be trivial as a holomorphic vector bundle in either sense, both of equations (13.35, 13.36) will generically have two linearly independent solutions, i.e. solutions for which (13.37) is non-zero. However, for exceptional two-surfaces, $S^{A'}$ will fail to be holomorphically trivial for one or other definition; solutions of the corresponding one of (13.35, 13.36) will be proportional as spinor fields and it will not be possible to define (non-zero) $\epsilon_{A'B'}$ by (13.37). For such an exceptional S, one of the constructions will fail. From (13.35), $S^{A'}$ will be exceptional if ρ' is zero on S, since then $\lambda^{0'}$ will be zero and all solutions will be proportional to $\iota^{A'}$; likewise from (13.36), $S^{A'}$ will be exceptional for the other definition if ρ is zero. These conditions for exceptionality are sufficient, but they are not necessary.

While exceptional two-surfaces present a problem, this problem corresponds to the existence of 'extra solutions' to the two-surface twistor equation in the original Penrose construction. There is no equivalent in the Dougan–Mason construction to the contorted/non-contorted dichotomy and the difficulties which it brings to the Penrose construction.

The integral (13.34) defines a momentum-vector $P^{A'}{}_A$ according to

$$P^{A'}{}_A = \frac{i}{4\pi G}\int_S \{\lambda_{(A'}^{A'}\nabla_{B')A}\bar{\lambda}_A^A - \bar{\lambda}_A^A\nabla_{A(A'}\lambda_{B')}^{A'}\}d\sigma^{A'B'} \tag{13.38}$$

and we use the ϵ-spinor together with its complex conjugate to calculate the norm of the momentum, i.e. the mass.

This then is the Dougan–Mason definition of quasi-local momentum and mass. Strictly speaking, it is a pair of definitions depending on whether one uses V or $\tilde{V}$, (13.35) or (13.36). What properties do these definitions have?

- First of all, in flat space, solutions of either of equations (13.35, 13.36) are necessarily constant on S, so that the momentum defined by (13.38) necessarily vanishes for any 2-surface S in flat space.

- By extending the spinor fields $\lambda^{A'}$ to a 3-surface Σ spanning S in a suitable way, it is possible to show that (13.38) gives the correct definition of momentum in linear theory.

- At spatial infinity one recovers the ADM momentum using either of equations (13.35, 13.36), but if one takes a two-surface S tending to null infinity then there is a difference between the two choices. Specifically, going to future-null-infinity one must use the definition based on (13.35), and one then recovers the Bondi momentum; the integral using (13.36) diverges towards future-null-infinity; the converse holds on the approach to past-null-infinity.

- For small spheres in the presence of matter, one again has 'volume times momentum density' for the momentum vector. In vacuum one obtains at $O(r^5)$ a future-pointing time-like vector constructed from the Bach tensor. At $O(r^6)$ one begins to see a difference between the use of V and $\tilde{V}$.

- The most striking feature of this definition is that it can be used to prove a positivity result:

If the Dominant Energy Condition (see e.g. Penrose and Rindler 1984) holds on a 3-surface Σ spanning the 2-surface S and if at S the spin-coefficient ρ' satisfies $\rho' \geq 0$, $\rho' \neq 0$ somewhere on S (respectively $\rho \leq 0$, $\rho \neq 0$ somewhere) then the momentum defined by (13.38) using (13.35) (repectively (13.36)) is time-like and future-pointing.

The condition on the spin-coefficients is a condition of *convexity* on the two-surface S; if $\rho' \geq 0$ then the area is non-decreasing at all points in the past-pointing, out-going null direction; if $\rho \leq 0$ then the area is non-decreasing at all points in the future-pointing, out-going null direction. Provided one of these two conditions holds, one of the two definitions of momenta is time-like and future-pointing. The proof involves a modification of the Witten argument (Witten 1981) using the spinors $\lambda^{A'}$ at S to provide boundary-values for the Witten equation on Σ (Dougan 1991; Dougan and Mason 1991).

This theorem makes the Dougan–Mason definition a particularly interesting one.

Exercises 13

a) Calculate the weights of the spin coefficients (13.17) and show that the operators (13.20) have well-defined weights.

b) Show that quantities of weight (p, q) on the sphere S are sections of a line bundle which may be identified with $\mathcal{O}(q-p)$ (see exercise 3g). Deduce that weighted functions can only be constant on S if $p = q$.

Chapter 14

Functionals on Zero Rest Mass Fields

Given the powerful constructions of chapters 8–10, which show how the Penrose transform describes zero rest mass fields and neatly geometrises the concept of positive frequency, it is natural to seek a twistor description of interactions. One might expect such a description to arise out of a twistor dynamical principle, replacing the usual space-time Lagrangian, and indeed the early work (Penrose and MacCallum 1972) proceeded in just this way, the resulting twistor integrals being referred to as *twistor diagrams*. Twistor diagrams share with Feynmann diagrams a simple combinatorial structure, but are not supposed to share their divergences! The analogy between twistor diagrams and Feynmann diagrams has been extensively explored, for which see Hodges (1990) and references therein. In this work twistor diagrams are regarded as determining integrals of holomorphic functions over compact contours, and it is only fairly recently that the diagrams have been reinterpreted as functionals on the cohomology *classes* representing the fields. This latter question, which is reviewed in Huggett (1990), is our subject here, and in order to discuss it adequately we will need to draw on some more cohomology theory. We start, however, by making a relatively simple observation on the geometry of products of twistor spaces.

In the product space $\{(Z^\alpha, W_\alpha) \in \mathbf{PT} \times \mathbf{PT}^* : Z^\alpha W_\alpha \neq 0\}$ the following holomorphic differential forms arise quite naturally. Let $DZ = \epsilon_{\alpha\beta\delta\gamma} Z^\alpha dZ^\beta dZ^\delta dZ^\gamma$, where $\epsilon_{\alpha\beta\delta\gamma}$ is as in (7.7), define DW similarly and let DZW be their wedge product. Then for each homogeneity we have the form

$$\frac{DZW}{(Z^\alpha W_\alpha)^n}. \tag{14.1}$$

These forms can be used as kernels for integrating products of homogeneous functions $f(Z)$ and $g(W)$ which are holomorphic on some parts of **PT** and **PT***. Indeed, given such an f and g, both homogeneous of degree -2, we can evaluate

$$\oint \frac{f(Z)g(W)DZW}{(Z^\alpha W_\alpha)^2}. \tag{14.2}$$

Remarkably, this integral has the following simple interpretation in space-time. Regard the functions f and g as representative cocycles for elements of first cohomology groups, and let ψ and χ be the scalar fields corresponding via (10.1,10.2) to these cohomology elements. Then (14.2) is equal to the usual (Gross 1964) Hermitian symmetric positive definite conformally invariant norm, which we refer to as the *scalar product* between these fields.

One may imagine generalising this to interactions involving more than two fields and more than two twistor spaces (or their duals) in the product. The kernels are still combinations of forms like (14.1) (perhaps including $I_{\alpha\beta}$), but they quickly become cumbersome to write out explicitly. It was partly for this reason, but more to emulate the Feynman diagrams of quantum field theory, that the twistor diagram notation was devised to represent these kernels. We are not going to describe twistor diagrams here; we simply observe that their theory aims to provide a (finite) twistor description of interactions between fields (z.r.m.s in the first instance). If a given such interaction involves a number of fields, then there would be a twistor diagram representing a kernel against which to integrate the product of representative cocycles for these fields. The diagram should also prescribe in some way a (compact) contour over which to perform this integration, although this is often quite problematic. Indeed a good part of the work in establishing that (14.2) is the scalar product comes from the identification of this contour. Even when such a contour has been identified, there remains the question of whether it defines a functional on the cohomology *classes* and not just the cocycles. (It also remains to demonstrate the Hermitian symmetry and positive definiteness in the twistor picture. See Baston and Eastwood (1989).)

In this chapter we will address the problem of finding *all* functionals on a given collection of z.r.m.s, these functionals all corresponding to a given kernel (which arises from a specified twistor diagram, say). Before we can do this though we will need some more cohomological theory. For a slightly more leisurely account, see Huggett and Singer (1991).

The reader will probably already be aware that there are various cohomology theories; de Rham cohomology is well known, and in chapter 9 we introduced Čech cohomology. Here we introduce Dolbeault cohomology and discuss its relationship with the de Rham and Čech theories. Before that, however, let us briefly review the de Rham cohomology.

On a (paracompact, Hausdorff, C^∞) manifold M of dimension n we define $\Omega^p(M)$ to be the group of C^∞ differential forms of degree p, and we let d denote the usual exterior derivative. Then there is a *complex* (i.e. a sequence of groups and group homorphisms such that 'd^2' $= 0$):

$$\Omega^0(M) \xrightarrow{d} \Omega^1(M) \xrightarrow{d} \cdots \xrightarrow{d} \Omega^p(M) \xrightarrow{d} \cdots \xrightarrow{d} \Omega^n(M) \tag{14.3}$$

and the de Rham cohomology groups are defined by

$$H^p(M) \quad = \quad \frac{\{\omega \in \Omega^p(M) : d\omega = 0\}}{d\Omega^{p-1}(M)}$$

$$(= \quad \text{closed } p\text{-forms modulo exact } p\text{-forms}).$$

In fact it can be proved fairly readily that these cohomology groups are the same as the Čech groups in which the sheaf is the constant sheaf **R**:

$$H^*(M) \cong \check{H}^*(M; \mathbf{R}).$$

If our manifold M happens to be a complex manifold (of *complex* dimension n) the groups $\Omega^p(M)$ have extra structure. This is most familiar when $p = 1$. Then

$$\Omega^1(M) = \Omega^{1,0}(M) \oplus \Omega^{0,1}(M)$$

where in local complex coordinates z_i the group $\Omega^{1,0}(M)$ consists of forms like

$$\sum_i f_i dz_i$$

while $\Omega^{0,1}(M)$ has elements

$$\sum_i g_i d\bar{z}_i.$$

For a general p we have

$$\Omega^p = \bigoplus_{i=0}^{p} \Omega^{p-i,i}(M)$$

where $\Omega^{r,s}(M)$ consists of forms 'of type (r, s)':

$$\sum_{\substack{j_1\cdots j_r \\ k_1\cdots k_s}} f_{j_1\ldots j_r k_1 \ldots k_s} dz_{j_1} \wedge \ldots \wedge dz_{j_r} \wedge d\bar{z}_{k_1} \wedge \ldots \wedge d\bar{z}_{k_s}.$$

The derivative operator d also decomposes:

$$d = \partial + \overline{\partial}$$

which (again) is most familiar when applied to functions in our local coordinates z_i:

$$df = \sum_{i=1}^{n} [\frac{\partial f}{\partial z_i} dz_i + \frac{\partial f}{\partial \overline{z}_i} d\overline{z}_i].$$

Here $d : \Omega^0(M) \rightarrow \Omega^1(M)$ has split into

$$\partial \quad : \quad \Omega^0(M) \rightarrow \Omega^{1,0}(M)$$
$$\overline{\partial} \quad : \quad \Omega^0(M) \rightarrow \Omega^{0,1}(M).$$

The induced decomposition of the action of d on (r, s) forms is obvious (but messy to write out explicitly!). It is clear, though, that we get

$$\partial \quad : \quad \Omega^{r,s}(M) \rightarrow \Omega^{r+1,s}(M)$$
$$\overline{\partial} \quad : \quad \Omega^{r,s}(M) \rightarrow \Omega^{r,s+1}(M).$$

Furthermore,

$$d^2 = (\partial + \overline{\partial})^2 = \partial^2 + \partial\overline{\partial} + \overline{\partial}\partial + \overline{\partial}^2 = 0$$

and each of the terms $\partial^2, \partial\overline{\partial} + \overline{\partial}\partial$ and $\overline{\partial}^2$ maps into a different direct summand, so in particular

$$\overline{\partial}^2 = 0.$$

We can therefore define, for any r, the *Dolbeault complex*

$$\Omega^{r,0}(M) \xrightarrow{\overline{\partial}} \Omega^{r,1}(M) \xrightarrow{\overline{\partial}} \cdots \xrightarrow{\overline{\partial}} \Omega^{r,s}(M) \xrightarrow{\overline{\partial}} \cdots \xrightarrow{\overline{\partial}} \Omega^{r,n}(M) \qquad (14.4)$$

and the Dolbeault cohomology groups are defined by

$$H_{\overline{\partial}}^{r,s}(M) = \frac{\{\omega \in \Omega^{r,s}(M) : \overline{\partial}\omega = 0\}}{\overline{\partial}\Omega^{r,s-1}(M)}.$$

(The alert reader will recall these definitions from exercise 9d.) It can quite easily be shown that

$$H_{\overline{\partial}}^{r,s}(M) = \check{H}^s(M; \Omega^r).$$

which gives us a very useful relationship between Dolbeault and Čech cohomology.

Let us return to the de Rham cohomology groups for a moment and recall the pairing between $H^p(M)$ and $H_p(M)$ given, simply, by integration.

Poincaré duality guarantees that this pairing is non-degenerate, and in the finite-dimensional case we have in fact

$$H^p(M) \cong H_p(M)^*.$$

There is another class of linear functionals on $H^p(M)$, though. We can consider the group $\Omega_c^p(M)$ of p-forms with *compact support*, and define

$$H_c^p(M)$$

in the obvious way (noting that $d : \Omega_c^p \to \Omega_c^{p+1}$). If $\omega \in \Omega^p(M)$ and $\eta \in \Omega_c^{n-p}(M)$ then

$$\int_M \omega \wedge \eta$$

is well defined and only depends on the cohomology classes of ω and η. So

$$H_c^{n-p}(M)$$

is a space of linear functionals on

$$H^p(M).$$

Indeed, it can be shown that

$$H_c^{n-p}(M) \cong H_p(M).$$

We can do just the same for the Dolbeault cohomology groups in the case when M is a complex manifold (of *complex* dimension n). Given

$$[\alpha] \in H^{r,s}(M)$$
$$[\beta] \in H_c^{n-r,n-s}(M)$$

we have a pairing given by integration

$$\int_M \alpha \wedge \beta.$$

We can see how this generalises our usual integration by putting $r = n$. Then we have

$$
\begin{array}{ccc}
H^{n,s}(M) & \to & H^{n+s}(M) \\
\otimes & & \otimes \\
H_c^{0,n-s}(M) & \leftarrow & H_c^{n-s}(M) \;\cong H_{n+s}(M) \\
\downarrow \;\; \text{pairing} & & \downarrow \;\; \text{usual} \\
\mathbb{C} \;\; \text{with a} & & \mathbb{C} \;\; \text{integration} \\
\text{compact} & & \\
\text{Dolbeault} & & \\
\text{form} & &
\end{array}
\qquad (14.5)
$$

In fact we will need our Dolbeault cohomology to have coefficients in sheaves of germs of holomorphic functions of various sorts. Let such a sheaf be denoted by $\mathcal{S}$. Then we tensor each group in the complex (14.4) and take cohomology (with respect to $\overline{\partial}$) to obtain

$$H^{r,s}(M; \mathcal{S})$$

and in the compactly supported case

$$H_c^{r,s}(M; \mathcal{S}).$$

We now have a pairing between

$$H^{r,s}(M; \mathcal{S}) \text{ and } H_c^{n-r,n-s}(M; \mathcal{S}^*)$$

where $\mathcal{S}^*$ is the dual sheaf to $\mathcal{S}$. (For example, $\mathcal{O}(-n)$ is dual to $\mathcal{O}(n)$.)

We need one last refinement of all this theory before we can actually use it, namely relative cohomology. Relative de Rham cohomology comes from the complex whose pth term is

$$\Omega^p(M, U) = \Omega^p(M) \oplus \Omega^{p-1}(U)$$

(where U is an open subset of M) with the differential

$$d(\alpha, \beta) = (d\alpha, \alpha|_U - d\beta)$$

mapping $\Omega^p(M, U)$ to $\Omega^{p+1}(M, U)$ (and satisfying $d^2 = 0$). We have the short exact sequence

$$\begin{array}{ccccccccc}
0 \to & \Omega^{p-1}(U) & \to & \Omega^p(M, U) & \to & \Omega^p(M) & \to 0 \\
& \beta & \to & (0, \beta) & & & \\
& & & (\alpha, \beta) & \to & \alpha &
\end{array}$$

and the corresponding long exact sequence

$$\cdots \to H^{p-1}(U) \to H^p(M, U) \to H^p(M) \to H^p(U) \to \cdots$$

called the *relative exact sequence*, in which the connecting map is restriction to U.

Relative Dolbeault cohomology comes from the complex whose sth term is

$$\Omega^{r,s}(M, U; \mathcal{S}) = \Omega^{r,s}(M, \mathcal{S}) \oplus \Omega^{r,s-1}(U, \mathcal{S})$$

with the differential

$$\overline{\partial}(\alpha, \beta) = (\overline{\partial}\alpha, \overline{\partial}\beta + (-1)^s \alpha|_U),$$

and we have the analogous short exact sequence of sheaves and long exact sequence of relative cohomology groups.

For relative Dolbeault cohomology with compact support, we define

$$\Omega_c^{r,s}(M, U; \mathcal{S})$$

from the exact sequence

$$0 \to \Omega_c^{r,s}(U, \mathcal{S}) \to \Omega_c^{r,s}(M, \mathcal{S}) \to \Omega_c^{r,s}(M, U; \mathcal{S}) \to 0.$$
$$\text{extend}$$
$$\text{by zero}$$

Then the qth cohomology group of the complex

$$\Omega_c^{r,*}(M, U; \mathcal{S}), \bar{\partial}$$

is $H^{r,q}(M, U; \mathcal{S})$. Again there is a relative long exact sequence, and this time the connecting map is induced by $\bar{\partial}$.

Now, at last, consider an interaction between m zero rest mass fields. Let U_j ($j = 1, \ldots, m$) be any open subset of P_j (which is **PT** or **PT***) swept out by projective lines corresponding to points in a convex set V_j in Minkowski space. Then the theorems of chapter 10 (which were given for the case $U = \mathbf{PT}^+$, $V = \mathbf{CM}^+$) apply:

$$H^1(U_j; \mathcal{O}(-n-2)) \cong \{\text{massless free fields of helicity } \frac{n}{2}$$
$$\text{holomorphic on } V_j\}.$$

We will also adopt the notation $F_j = P_j - U_j$, L_j for a projective line contained in P_j, $\Pi = P_1 \times \ldots \times P_m$, $F = F_1 \times \ldots \times F_m$, and $\Lambda = L_1 \times \ldots \times L_m$.

We will be studying the space of functionals on the tensor product

$$H^1(U_1; \mathcal{O}(r_1)) \otimes \cdots \otimes H^1(U_m; \mathcal{O}(r_m)) \qquad (14.6)$$

with the object of picking out those determined by a given twistor diagram. This tensor product is *much* more tractable if we use relative H^2s to represent our fields, instead of H^1s. We *can* do this, because of the following relative long exact sequence:

$$H^1(P_j; \mathcal{O}(r_j)) \to H^1(U_j; \mathcal{O}(r_j)) \to H^2(P_j, U_j; \mathcal{O}(r_j)) \to H^1(P_j; \mathcal{O}(r_j))$$
$$= 0 \qquad\qquad\qquad\qquad\qquad\qquad\qquad\qquad\qquad = 0$$

So

$$H^1(U_j; \mathcal{O}(r_j)) \cong H^2(P_j, U_j; \mathcal{O}(r_j)) \quad \forall j.$$

Theorem

The tensor product of cohomology groups (14.6) is isomorphic to

$$H^{2m}(\Pi, \Pi - F; \mathcal{O}(\underline{r}))$$

where $\mathcal{O}(\underline{r}) = \mathcal{O}(r_1) \times \ldots \times \mathcal{O}(r_m)$.

Proof

We sketch the proof in the case $m = 2$: the scalar product. The proof
for general m is only notationally more fiddly. We have, by the Künneth
Theorem (Spanier 1989) for relative cohomology,

$$H^4(\Pi, \Pi - F; \mathcal{O}(\underline{r})) = \bigoplus_{\substack{p,q \\ p+q=4}} H^p(P_1, U_1; \mathcal{O}(r_1)) \otimes H^q(P_2, U_2; \mathcal{O}(r_2)).$$

But we can easily see from the relative long exact sequence and the geometric properties of U_j that

$$
\begin{aligned}
H^p(P_j, U_j; \mathcal{O}(r_j)) &= 0 \text{ for } p < 2. \\
H^4(\Pi, \Pi - F; \mathcal{O}(\underline{r})) &= H^2(P_1, U_1; \mathcal{O}(r_1)) \otimes H^2(P_2, U_2; \mathcal{O}(r_2)) \\
&= H^1(U_1; \mathcal{O}(r_1)) \otimes H^1(U_2; \mathcal{O}(r_2))
\end{aligned}
$$

as required.

$\square$

Our next task is to represent, in cohomological terms, the differential
form specified by the interior of the diagram. In the scalar product case we
have the element $(W_\alpha Z^\alpha)^{-r} DZW$ of

$$H^{6,0}(\Pi - \Sigma; \mathcal{O}(-\underline{r}))$$

where $\Sigma = \{W_\alpha Z^\alpha = 0\}$ and $\underline{r} = (r, r)$. In general we will allow ourselves
more freedom than this, and think of the diagram as determining a kernel

$$h \in H^{3m,q}(\Pi - \Sigma; \mathcal{O}(-\underline{r})).$$

The space of *all* functionals on $H^{2m}(\Pi, \Pi - F; \mathcal{O}(\underline{r}))$ can be deduced
from the following commutative diagram (for $r = 0$, $s = 2m$ and $n = 3m$),
in which the complex dimension of M is $3m$:

$$
\begin{array}{ccccccc}
\rightarrow & H^{r,s}(M; \mathcal{S}) & \rightarrow & H^{r,s}(U; \mathcal{S}) & \rightarrow & H^{r,s}(M, U; \mathcal{S}) & \rightarrow \\
& \otimes & & \otimes & & \otimes & \\
\leftarrow & H_c^{n-r,n-s}(M; \mathcal{S}^*) & \leftarrow & H_c^{n-r,n-s}(U; \mathcal{S}^*) & \leftarrow & H_c^{n-r,n-s}(M, U; \mathcal{S}^*) & \leftarrow \\
& \downarrow & & \downarrow & & \downarrow & \\
& \mathbb{C} & & \mathbb{C} & & \mathbb{C} &
\end{array}
$$

We also use the fact that the first two pairings (which are those we saw earlier) have the property that *any* continuous linear functional on $H^{r,s}(U; \mathcal{S})$ (say) can be obtained by pairing it with an element of $H^{n-r,n-s}(U; \mathcal{S}^*)$. So our space of functionals is

$$H_c^{3m,m}(\Pi, \Pi - F; \mathcal{O}(-\underline{r}))$$

and we need to see how our kernel h picks out one (or a small finite number) of these. We first use the open inclusion

$$i : \Pi - \Sigma \to \Pi$$

to cut out Σ:

$$i^* : H_c^{3m,m}(\Pi - \Sigma, \Pi - \Sigma \cup F; \mathcal{O}(-\underline{r})) \to H_c^{3m,m}(\Pi, \Pi - F; \mathcal{O}(-\underline{r})).$$

and then we note that cupping with our kernel h defines a map

$$\cup h : H_c^{0,m-q}(\Pi - \Sigma, \Pi - \Sigma \cup F) \to H_c^{3m,m}(\Pi - \Sigma, \Pi - \Sigma \cup F; \mathcal{O}(-\underline{r})).$$

So our functionals are determined by elements

$$\alpha \in H_c^{0,m-q}(\Pi - \Sigma, \Pi - \Sigma \cup F).$$

But these elements are more than just contours. They can be thought of as 'function-contours', i.e. contours with germs of holomorphic functions defined on them, as was described very neatly by Jozsa (in Hughston and Ward 1979). In order to insist that our functional arises from a straightforward contour, we look back at (14.5) and see that we require α to be in the image of the map

$$H_{5m+q}(\Pi - \Sigma, \Pi - \Sigma \cup F) \to H_c^{0,m-q}(\Pi - \Sigma, \Pi - \Sigma \cup F),$$

which embeds the ordinary contours into the space of function-contours.

In the scalar product case $q = 0$ and $m = 2$ so we are looking for elements of

$$H_{10}(\Pi - \Sigma, \Pi - \Sigma \cup F) \tag{14.7}$$

and if our fields are elementary states (see exercise 8d) we have $F = \Lambda$. More generally, suppose only that $(\Pi - \Sigma, \Pi - \Sigma \cup F)$ is homotopic to $(\Pi - \Sigma, \Pi - \Sigma \cup \Lambda)$. Then (14.7) becomes

$$H_{10}(\Pi - \Sigma, \Pi - \Sigma \cup \Lambda)$$

which is isomorphic to

$$H_2(\Lambda - \Sigma) \tag{14.8}$$

by the Thom isomorphism theorem (Bott and Tu (1982)). This latter group has one generator, so the scalar product diagram determines one functional, and this functional can be shown to coincide with the contour integral (14.2), as we would hope!

In general our functionals would be given by elements of

$$H_{m+q}(\Lambda - \Sigma) \tag{14.9}$$

for some q and Σ, or if the Thom isomorphism theorem were unavailable,

$$H_{5m+q}(\Pi - \Sigma, \Pi - \Sigma \cup F).$$

Finally we make a few comments on how to tell if a contour is cohomological. The question is this: a twistor diagram with m elementary states has at least $2m$ poles in the integrand. A *practical* way of proceeding with the evaluation of such a diagram is to take residues at each of these poles, and *then* to tackle the remainder of the integral. Does this correspond to a cohomological evaluation, and are *all* cohomological evaluations captured in this way?

Having taken all these residues, one would be left seeking a contour in $H_{m+q}(\Lambda - \Sigma)$, so it seems from (14.9) that both answers are yes. Well, they are, but one has to be a little more careful: taking $2m$ residues is dual to applying $2m$ *cobord* maps. (A cobord map, applied to a contour lying in a real codimension 2 submanifold S, replaces each point x of the contour by an S^1 lying in the normal space to S at x.) On the other hand, the Thom isomorphism we used in (14.8) involves the application of one 'big' cobord map. (Here the submanifold is of real codimension 8 and the Thom isomorphism replaces each point x in the contour by a normal 8-dimensional disc. This, followed by a boundary map, yields an S^7 above each point x in the original contour, which is our 'big' cobord map.) An interesting technical lemma (Huggett and Singer 1990) allows us to show that these procedures coincide for contours without boundary, so that *any* such cohomological evaluation can be implemented by taking these $2m$ residues first. The remaining question concerns contours with boundary.

Chapter 15

Further Developments and Conclusions

In this book, our aim has been to provide a brief introduction to twistor theory, sketching the mathematical background and indicating where it makes contact with the physics of space-time. It is possible to carry on from here in directions which we may characterise as *inner* or *outer*, and *mathematical* or *physical*.

By *inner* we mean developments which make rigorous and treat more thoroughly what has been treated informally here. For spinor theory the main reference is the two volume *Spinors and space-time* of Penrose and Rindler (1984, 1986). This is a mighty compendium of results for spinors in flat and curved space-time, and has more detail, in volume 2, on the quasi-local mass construction described in chapter 13. For more on sheaf cohomology and the twistor theory of massless fields and the active, curved-twistor-space constructions of chapter 11 see *Twistor Geometry and Field Theory* (Ward and Wells 1990).

By *outer* we mean developments that go beyond what has been covered here. Other parts of the theory of z.r.m. fields have been given a twistor description, and the formalism of chapters 8 to 10 has been much extended. Hyperfunction z.r.m. fields are discussed in Bailey et al. (1982) and fields with sources, which can be described by relative cohomology, by Bailey (1985). This material is reviewed in the article of Bailey and Singer in Bailey and Baston (1990).

There is a large body of theory concerned with *twistor diagrams* which are the counterpart of Feynman diagrams, in that twistor diagrams define contour integrals which represent scattering amplitudes for various scattering processes involving z.r.m. fields. In chapter 14 we saw how contour integrals of this type may be described in terms of sheaf cohomology. For the theory of the twistor diagrams themselves, see Penrose and MacCallum (1973); Sparling (1975); Hodges (1982, 1983a, 1983b, 1985a, 1985b) and the articles by Hodges and by Huggett in Bailey and Baston (1990). This theory at present lacks a twistor theoretic generating principle (such as Feynman diagrams have in space-time Lagrangians). An interesting proposal, though, is the twistor conformal field theory for four space-time dimensions (Hodges et al. 1989; Singer 1990; Huggett 1992). The Riemann surfaces and holomorphic functions of the more usual two-dimensional conformal field theory are here replaced by (a class of) complex three-manifolds and holomorphic sheaf cohomology, with the aim of generating interaction amplitudes between the space-time fields represented by the cohomology elements.

Massive fields have a twistor contour-integral formulation (Penrose and MacCallum 1973) with an analogous cohomological apparatus (Hughston and Hurd 1981; Eastwood 1981). A geometrical view of the spin-and-statistics theorem can be given from this point of view.

In chapter 7 we saw that a null geodesic is represented by a null twistor. A time-like geodesic can be represented by a combination of twistors, which leads to the representation of a massive spinning particle twistorially (Penrose and MacCallum 1973; Tod and Perjes 1976; Tod 1977). This combination has an internal symmetry group which is typically a semi-direct product involving $SU(n)$. From representations of these *twistor internal symmetry groups* it is possible to build up a twistorial classification of elementary particles which is similar to, but differs in important respects from, the standard classifications (Hughston 1979).

Many examples have been given of the constructions of chapter 11 solving the a.s.d. Yang–Mills equations (Atiyah and Ward 1977; Atiyah et al. 1978; Corrigan et al. 1978; Woodhouse 1983; see also Ward and Wells 1989) and of the construction of chapter 12 solving the a.s.d. vacuum equations (Curtis et al. 1978; Hitchin 1979; Ward 1978; Tod and Ward 1979). The construction of chapter 11 has been generalised to give monopoles (Ward 1981; Hitchin 1982, 1983) while the construction of chapter 12 has been generalised to give a.s.d. Einstein (but non-vacuum) spaces (Ward 1980; LeBrun 1982) and scalar-flat, Kähler manifolds (LeBrun 1991).

A development from the study of a.s.d. Yang–Mills fields has been the study, in twistor terms, of integrable equations (see the article of Ward in Bailey and Baston 1990). Many of the known, completely-integrable p.d.e.s turn out to be reductions of the a.s.d. Yang–Mills equations, and the Lax-pair formulations associated with integrable equations often turn out to be interpretable as curvature conditions on a connection like (11.16).

(By a reduction we mean that the Yang–Mills potential is assumed to be invariant under a subgroup of the conformal group of Minkowski space, possibly with some extra assumptions on the form of the potential.) There are a few known integrable equations which have not been obtained from a.s.d. Yang–Mills, and it has been suggested by Mason (recorded in Tod 1992) that these will arise from the a.s.d. Einstein equations. According to this conjecture, the non-linear graviton is a *universal integrable system* from which all others are obtained by reduction.

The real, Lorentzian-signature Einstein vacuum equations for stationary, axisymmetric space-times are equivalent to a certain reduction of the a.s.d. Yang–Mills equations. This fact has been exploited by Woodhouse and Mason (1988) and by Fletcher and Woodhouse (in Bailey and Baston 1990) to give a a twistor construction of such space-times in terms of holomorphic bundles over non-Hausdorff Riemann surfaces. Their formalism provides a geometrical way of understanding and unifying the considerable literature on this class of solutions of Einstein's equations.

One type of reduction of the a.s.d. Einstein equations is the Einstein–Weyl equations. These are a kind of 3-dimensional conformally invariant generalisation of the Einstein equations which therefore provide another set of non-linear p.d.e.s with a geometrical interpretation which is solvable by a twistor construction (Jones and Tod 1985; Tod 1990b).

Returning to a.s.d. Einstein spaces, since these can be real for Riemann signature, the Riemannian case has been of particular interest to mathematicians. There is a large literature on positive-definite twistor theory following the original paper of Atiyah et al. (1978) (see e.g. Besse 1987). A related development of great mathematical significance has been the study of the moduli-space of a.s.d. Yang–Mills fields on a given 4-manifold as a means of studying the topology of 4-manifolds (Donaldson and Kronheimer 1990).

A twistor characterisation of the full, as opposed to s.d. or a.s.d., Yang–Mills equations has been given (Isenberg et al. 1978; Witten 1978). The construction uses bundles over *ambi-twistor space*, a space of pairs of twistors which can be identified with the space of null geodesics in Minkowski space. A similar thing has been done for the Yang–Mills–Dirac system (Henkin and Manin 1980; Manin 1988).

Super-symmetric extensions of twistor theory, of the non-linear graviton and the Einstein–Weyl equations have been given by Merkulov (1991, 1992a,b).

Other ways of describing twistors in curved spaces are discussed in Penrose and MacCallum (1973); Penrose (1975); Penrose and Ward (1980). In particular, there is a close connection with Newman's H space (Newman 1976; Hansen et al. 1978). The holy grail of general relativity is a construction of the general, real, Lorentzian-signature solution of the vacuum equations. A twistor characterisation of these has been given in terms of deformations of the complex structure of ambi-twistor space (Baston and Mason 1987; LeBrun in Bailey and Baston 1990); this is a characterisation rather than a solution since the deformations are not given in terms of free data. A more recent attempt to find the general solution of the vacuum equations has been to try to define twistors as charges of spin-$\frac{3}{2}$ fields, since the vacuum equations are the integrability conditions for such fields to exist (in the sense of 'potentials modulo gauge' (Penrose 1992)).

There is an almost separate tradition in twistor theory which invloves the approach to complex manifolds outlined in exercises 3h and 9d. Holomorphic objects are thought of as constructed from real differentiable objects satisfying Cauchy–Riemann equations (Woodhouse 1985). In this tradition, cohomology classes for example are globally defined differential forms rather than functions defined on overlaps. Likewise the bundles and fibrations of chapters 11 and 12 are defined in terms of modified Cauchy–Riemann equations. In particular, Newman's good cut equation (Newman 1976; Tod 1982) can be interpreted in this way.

Other material not previously mentioned can be found in the book of commissioned reviews (Bailey and Baston 1990), in the collected reprints from Twistor Newsletter (Mason and Hughston 1990), in the book of Baston and Eastwood (1989), and in the proceedings of the 1993 Twistor Theory conference in Devon (Huggett 1994).

In this book, one of our aims has been to convey some sense of the variety of twistor theory. Initially the theory is one of space-time geometry with an emphasis on conformal invariance and the group isomorphisms which introduce the complex numbers:

$$SL(2,\mathbf{C}) \overset{2-1}{\to} O_+(1,3)$$

$$SU(2,2) \overset{2-1}{\to} O(2,4) \overset{2-1}{\to} C(1,3).$$

The discussion of z.r.m. fields leads us into the complex calculations of sheaf cohomology and gives a first instance of the disappearance of field equations into complex analyticity.

The active constructions of chapters 11 and 12 involve more elaborate complex manifold theory and, as recompense, solve *non-linear* field equations by complex analyticity.

The further developments which we have just sketched here lead on into a massless quantum field theory, classification schemes for elementary particles, geometrisations of other branches of physics and the possible solution of a completely classical and long-standing problem in general relativity.

Through this wide-ranging activity, the unifying thread is the central importance of holomorphic and geometrical ideas in space-time physics.

Chapter 16

Hints, Solutions and Notes to the Exercises

Chapter 2

a) For R_{abcd} in terms of S_{abcd}, consider $S_{a[bc]d}$.

b) For $\epsilon^{abcd}F_{ab}F_{cd} = 0 \Rightarrow F_{ab}$ simple, first look at components to prove that $\epsilon_a{}^{cde}F_{bc}F_{de} = \frac{1}{4}g_{ab}\epsilon^{fcde}F_{fc}F_{de}$.

d) You need $\mathcal{L}_X\epsilon_{ab}{}^{cd} = 0$; write it out and use the result of (c).

e) We have chosen to define the Lie derivative from the covariant derivative. An exercise like this then requires a calculation in tensor calculus. In a different, more abstract approach, the fact proved here would be elementary but then other facts would be more obscure.

Chapter 3

a) Show first that $t \in SL(2, \mathbf{C})$ has a unique *polar decomposition* $t = RU$ where R is a positive definite Hermitian matrix and U is in $SU(2)$.

d) The formula for R_{abcd} follows from repeated applications of the technique leading to (3.7).

e) See Field (1982) or Griffiths and Harris (1978).

g) For the line bundle $\mathcal{O}(-1)$ over $\mathbf{CP}^1$ the trivialisations are

$$t_0([\pi_{A'}], \lambda\pi_{A'}) = ([\pi_{A'}], \lambda\pi_{0'})$$

$$t_1([\pi_{A'}], \lambda \pi_{A'}) = ([\pi_{A'}], \lambda \pi_{1'}).$$

So $\mu_0 = \lambda \pi_{0'}$ and $\mu_1 = \lambda \pi_{1'}$, and hence the transition function f_{01} is $\frac{\pi_{0'}}{\pi_{1'}}$ as required. Local sections s_i satisfying the transition relation

$$s_0 = f_{01} s_1$$

have the property that

$$\frac{s_1}{\pi_{1'}} = \frac{s_0}{\pi_{0'}}.$$

Define

$$F(\pi_{A'}) = \begin{cases} \dfrac{s_1}{\pi_{1'}} & \text{where } \pi_{1'} \neq 0 \\[2mm] \dfrac{s_0}{\pi_{0'}} & \text{otherwise.} \end{cases}$$

Then F is homogeneous of degree -1.

h) For the last part, if V^a is a type $(1,0)$ vector-field then $V^{(a}\overline{V}^{b)}$ defines a (contravariant) conformal structure; conversely given a conformal structure and a real vector field U^a, one has a unique vector field which is '90° anti-clockwise' from U^a and the same length; this is $J_a{}^b U^a$.

Chapter 4

a) The problem is to find an example of the geometric situation illustrated in figure 4.1. One begins by imagining the rotation through 2π of an orthonormal frame at the point corresponding to the north pole in $\mathbf{CP}^1 = S^2$ and the origin in $\mathbf{C}$. One then seeks a sequence of paths on the S^2 beginning and ending at the north pole, with the first and last in the sequence being the zero path and the others gradually covering the S^2. Finally one considers the effect on the chosen orthonormal frame of being carried along these paths. For more details see Geroch (1968).

e) Differentiate again and use the Ricci identities (4.4, 4.5).

Chapter 5

b) See Penrose and Rindler (1984) section 6.8.

c) Writing $x^a(s)$ as x^a, and $x^a(0)$ as p^a, (5.6) becomes

$$x^a = \frac{p^a - sB^a p^2}{1 - 2sB_b p^b + s^2 B^2 p^2}.$$

Now calculate dx^a (for s and B^a constant). There will be a dp^a term, a $p_b dp^b$ term, and a $B_b dp^b$ term. Finally, calculate $dx_a dx^a$. By a miracle, only the $dp_a dp^a$ term survives, and with the correct factor!

d) $\begin{aligned}\mathcal{L}_X \Sigma &= \mathcal{L}_X(\omega^A \overline{\pi}_A + \overline{\omega}^{A'} \pi_{A'}) \\ &= \overline{\pi}_A \mathcal{L}_X \omega^A + \omega^A \mathcal{L}_X \overline{\pi}_A + \text{ complex conjugate} \\ &= \overline{\pi}_A X^b \nabla_b \omega^A - \overline{\pi}_A \phi^A{}_B \omega^B - \tfrac{k}{4} \overline{\pi}_A \omega^A \\ &\quad + \omega^A X^b \nabla_b \overline{\pi}_A + \omega^A \phi_A{}^B \overline{\pi}_B + \tfrac{k}{4} \omega^A \overline{\pi}_A + \text{ c.c.} \\ &= \overline{\pi}_A X^b \nabla_b \omega^A + \text{ c.c.} \\ &= -i X^{BB'} \overline{\pi}_B \overline{\pi}_{B'} + \text{ c.c.} \quad \text{from the twistor equation} \\ &= 0 \text{ because } X^b \text{ is real.} \end{aligned}$

e) After the coordinate change the metric becomes

$$\cosh^2 t \{ \operatorname{sech}^2 t \, dt^2 - [d\psi^2 + \sin^2 \psi(d\theta^2 + \sin^2 \theta d\phi^2)] \}$$

which is conformal to

$$d\tau^2 - [d\psi^2 + \sin^2 \psi(d\theta^2 + \sin^2 \theta d\phi^2)].$$

Now $T = \sinh t$ and $t = \log(\tan \frac{\tau}{2})$, so

$$T = -\cot \tau$$

and so $0 < \tau < \pi$ as required for the Einstein Static Universe.

Chapter 6

a) Necessity is easy. Sufficiency is a case of Frobenius' Theorem (Spivak 1975); the problem is to show that for any two vectors orthogonal to V^a, their Lie bracket is also orthogonal to V^a.

b) $\begin{aligned}l^a \text{ is h.s.o.} \quad &\Leftrightarrow \quad l_{[a} \nabla_b l_{c]} = 0 \quad \text{from part (a)} \\ &\Leftrightarrow \quad {}^*\{\nabla_{[a} l_{b]}\} l^a = 0 \quad \text{where } {}^* \text{ means the dual} \\ &\Leftrightarrow \quad l^a \{\nabla_{AB'} l_{BA'} - \nabla_{BA'} l_{AB'}\} = 0 \\ &\Leftrightarrow \quad o^A o^{A'} \nabla_{AB'} o_B o_{A'} = o^A o^{A'} \nabla_{BA'} o_A o_{B'} \\ &\Leftrightarrow \quad o^A D o_A = 0 \text{ and } \rho = \overline{\rho} \end{aligned}$

c) Solve $f = 0$ for o_A in terms of $x^{AA'}$. You get

$$\frac{o^1}{o^0} = \frac{x + iy}{z \pm \sqrt{x^2 + y^2 + z^2}}.$$

Now show that $l^a = o^A o^{A'}$, which depends on $x^{AA'}$, is coplanar with $t^{AA'}$ and $x^{AA'}$.

Chapter 7

a) Show that the eigenvalues of Σ are $+1$ (twice) and -1 (twice).

b)
$$
\begin{aligned}
\nabla_{AA'}\alpha_B\bar{\beta}_{B'} + \nabla_{BB'}\alpha_A\bar{\beta}_{A'} &= \alpha_B\nabla_{AA'}\bar{\beta}_{B'} + \bar{\beta}_{B'}\nabla_{AA'}\alpha_B \\
&\quad + \alpha_A\nabla_{BB'}\bar{\beta}_{A'} + \bar{\beta}_{A'}\nabla_{BB'}\alpha_A \\
&= -\alpha_B\nabla_{AB'}\bar{\beta}_{A'} - \bar{\beta}_{B'}\nabla_{BA'}\alpha_A \\
&\quad + \alpha_A\nabla_{BB'}\bar{\beta}_{A'} + \bar{\beta}_{A'}\nabla_{BB'}\alpha_A \\
&\qquad\qquad\qquad\qquad\text{from (7.1)} \\
&= (\alpha_A\nabla_{BB'} - \alpha_B\nabla_{AB'})\bar{\beta}_{A'} \\
&\quad + (\bar{\beta}_{A'}\nabla_{BB'} - \bar{\beta}_{B'}\nabla_{BA'})\alpha_A .
\end{aligned}
$$
Each term is now skew in both AB and $A'B'$.

c) Suppose $(\omega^A, \pi_{A'})$ is an arbitrary twistor on the line corresponding to $x^a - iy^a$. Then
$$
\omega^A = i(x^{AA'} - iy^{AA'})\pi_{A'} .
$$
Show that this is in $\mathbf{PT}^+ \Leftrightarrow y^{AA'}\bar{\pi}_A\pi_{A'} > 0$, so that the condition on y^a is that
$$
y^a p_a > 0
$$
for all null future-pointing p_a.

d)
$$
X^{[\alpha}P^{\beta\gamma]} = 0 \Rightarrow X^\alpha Q_{\beta\gamma}P^{\beta\gamma} + (X^\beta R^\gamma - X^\gamma R^\beta)Q_{\beta\gamma}S^\alpha
$$
$$
+ (X^\gamma S^\beta - X^\beta S^\gamma)Q_{\beta\gamma}R^\alpha = 0
$$
where $Q_{\beta\gamma}$ is any line skew to $P^{\beta\gamma}$ (so $Q_{\beta\gamma}P^{\beta\gamma} \neq 0$), and $P^{\beta\gamma} = R^{[\beta}S^{\gamma]}$. Thus X^α is a linear combination of R^α and S^α.

e) First show that $\frac{1}{2}\epsilon_{\alpha\beta\gamma\delta}P^{\gamma\delta}$ is skew and simple, so that
$$
\frac{1}{2}\epsilon_{\alpha\beta\gamma\delta}P^{\gamma\delta} = A_{[\alpha}B_{\beta]} .
$$

Next show that A_α and B_β are (distinct) planes through the line L defined by $P^{\gamma\delta}$. If R^α and S^β are any two (distinct) points on L then
$$
\overline{P}_{\alpha\beta} = \overline{R}_{[\alpha}\overline{S}_{\beta]} ,
$$
so L is real $\Leftrightarrow \overline{R}_{[\alpha}\overline{S}_{\beta]} = A_{[\alpha}B_{\beta]}$. Now it is easy to show that if L is real then R^α must be null. Conversely, if R^α and S^β (and any linear combination of them) are null then it is easy to show that $R^\alpha\overline{S}_\alpha = 0$ and hence that any plane through L is a linear combination of $\overline{R}_\alpha$ and $\overline{S}_\alpha$.

f) This double fibration is a special case of a *correspondence* between two complex manifolds Z and X given by a third complex manifold Y and two (surjective, maximal rank) mappings η and τ:

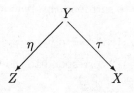

The pair (η, τ) must embed Y as a submanifold of $Z \times X$. The Penrose transform (described in chapter 10) can also be studied in this more general setting. See Baston and Eastwood (1989), for example.

g) Through any two points in **PT** there is a unique line. Now consider a point $x^{AA'}$ on the α-plane

$$A^A = ix^{AA'} A_{A'}$$

and on the β-plane

$$B^{A'} = -ix^{AA'} B_A.$$

It is immediate that $A^\alpha B_\alpha = 0$. The converse is much easier in the twistor picture: if $A^\alpha B_\alpha = 0$ there is a line L through A^α lying in B_α, so there is a point common to the α- and β-planes. In fact there is a 1-complex-parameter *pencil* of lines of this type, all meeting at A^α. Hence the result.

h) We refer back to (5.6) and (5.7), in which we considered the $O(2,4)$ null cone N. Instead of taking the space of generators, which yielded compactified Minkowski space, we now simply put $V = -1$, obtaining the de Sitter space described in exercise 5e. This hyperplane intersects all the generators of N except those for which $V = 0$. In **RP**^{5}, the $V = 0$ hyperplane is *not* tangent to the quadric Q (being the polar plane of the point $I^{\alpha\beta}$ *not* on Q).

Chapter 8

a) Put

$$f(Z^\alpha) = \frac{g(Z^\alpha)}{Z^\alpha A_\alpha}$$

where g is holomorphic and non-zero on $Z^\alpha A_\alpha = 0$. Then

$$\phi_{A' \ldots B'}(x) = \frac{1}{2\pi i} \oint_\gamma \pi_{A'} \ldots \pi_{B'} \rho_x \left(\frac{g}{Z^\alpha A_\alpha} \right) \pi_{C'} d\pi^{C'}$$

$$\rho_x \left(\frac{1}{Z^\alpha A_\alpha} \right) = \frac{1}{o^{A'} \pi_{A'}}, \quad \text{where } o^{A'}(x) = A^{A'} + ix^{AA'} A_A.$$

Now for a given x choose γ to surround this pole in $\pi_{A'}$ with a simple loop, and calculate

$$o^{A'}\phi_{A'...B'}(x).$$

In general, the cases $r = 0, 1, 2$ correspond to the types $N, 3, D$.

b) $\oint \pi_{A'} \ldots \pi_{B'} \rho_x (Z^\alpha A_\alpha f) \pi_{C'} d\pi^{C'}$

$\quad = \oint \pi_{A'} \ldots \pi_{B'} \alpha^{D'} \pi_{D'} \rho_x(f) \pi_{C'} d\pi^{C'}$, where $\alpha^{A'} = A^{A'} + ix^{AA'} A_A$

$\quad = \alpha^{D'} \phi_{A'...B'D'}$

and similarly for spin-raising.

c) See Penrose and MacCallum (1973) and Hughston and Hurd (1981).

d) Let $p^{A'} = P^{A'} + ix^{AA'}P_A$ etc., as in (8.2). Then the integral is

$$\frac{1}{2\pi i} \oint \frac{p^{A'}\pi_{A'}r^{B'}\pi_{B'} + q^{C'}\pi_{C'}s^{D'}\pi_{D'}}{(p^{F'}\pi_{F'}q^{G'}\pi_{G'})^2} \pi_{E'} d\pi^{E'}.$$

Choosing a local coordinate $z = \frac{p^{A'}\pi_{A'}}{q^{B'}\pi_{B'}}$ and calculating the residue at $z = 0$ (for which the 'epsilon identity' $\epsilon_{A[B}\epsilon_{CD]} = 0$ is needed), one obtains the field

$$\frac{p^{A'}r_{A'} + s^{B'}q_{B'}}{(p^{C'}q_{C'})^2}.$$

This is singular on the null cone of the point represented by $P_{[\alpha}Q_{\beta]}$, as required, and it is invariant under the transformations

$$r \mapsto r + \alpha p + \beta q$$
$$s \mapsto s - \beta p + \gamma q$$

where α, β and γ are complex parameters.

Chapter 9

b) No, because the groups of sections of a sheaf over open sets have the following property: if $f \in \mathcal{B}(U)$ and $g \in \mathcal{B}(V)$ are such that

$$f \mid_{U \cap V} = g \mid_{U \cap V}$$

then $\exists h \in \mathcal{B}(U \cup V)$ such that

$$h \mid_U = f \text{ and } h \mid_V = g.$$

This is one of the properties of a 'complete presheaf'; see Field (1982) chapter 6, especially Proposition 6.1.4.

c) Consider the mapping

$$r^* : \check{H}^1(\{U_i\}; \mathcal{S}) \to \check{H}^1(\{W_\alpha\}; \mathcal{S})$$

where $\{W_\alpha\}$ is a refinement of $\{U_i\}$ (with refining map r). Observe that the mapping

$$r^* : \check{C}^0(\{U_i\}; \mathcal{S}) \to \check{C}^0(\{W_\alpha\}; \mathcal{S})$$

is onto, and use the commutativity of r^* and δ.

d) For the second part of the exercise, first note that there is an isomorphism between $\mathcal{O}(-2)$ and Ω^1, the sheaf of germs of holomorphic 1-forms. Then use the short exact sequence

$$0 \to \Omega^1 \xrightarrow{i} \mathcal{E}^{1,0} \xrightarrow{\bar{\partial}} \mathcal{Z}^{1,1} \to 0,$$

where $\mathcal{Z}^{1,1}$ is the sheaf of germs of $\bar{\partial}$-closed (1,1) forms, to construct the following commutative diagram:

$$
\begin{array}{ccccccccc}
0 \to & \check{C}^0(\{U_i\}; \Omega^1) & \xrightarrow{i} & \check{C}^0(\{U_i\}; \mathcal{E}^{1,0}) & \xrightarrow{\bar{\partial}} & \check{C}^0(\{U_i\}; \mathcal{Z}^{1,1}) & \to 0 \\
& \downarrow \delta & & \downarrow \delta & & \downarrow \delta & \\
0 \to & \check{C}^1(\{U_i\}; \Omega^1) & \xrightarrow{i} & \check{C}^1(\{U_i\}; \mathcal{E}^{1,0}) & \xrightarrow{\bar{\partial}} & \check{C}^1(\{U_i\}; \mathcal{Z}^{1,1}) & \to 0
\end{array}
$$

Now the isomorphism

$$\check{H}^1(\{U_i\}; \Omega^1) \xrightarrow{\cong} H^1(\mathbf{P}^1; \Omega^1)$$

is effected by mapping the Čech class $[w]$ to the Dolbeault class $[\bar{\partial}(\delta^{-1}(iw))]$. The map δ^{-1} is multiplication by a partition of unity. If this cover $\{U_i\}$ is as in (9.4), then the appropriate partition of unity is

$$\left\{ \frac{|\pi_{0'}|^2}{|\pi_{0'}|^2 + |\pi_{1'}|^2}, \frac{|\pi_{1'}|^2}{|\pi_{0'}|^2 + |\pi_{1'}|^2} \right\}.$$

e) See Spanier (1989).

f) It is easiest to use a fine resolution of $\mathcal{S}$. Suppose $\mathcal{S} = \mathcal{O}$, for example, and use the resolution in exercise 9d. Then $C^p = \mathcal{E}^{0,p}$.

Chapter 12

b) This is a difficult exercise, and getting signs and factors right is a tricky business. If $\lambda = 0$, then a typical line L_x is given by

$$\hat{\omega}^A = \omega^A - i x^{AA'} \pi_{A'}. \tag{16.1}$$

Suppose the 'linearised deformed line' is given by adding terms of order λ to this. The main idea is to restrict the linearised transition relation to L_x and split it. Explicitly:

$$\rho_x \frac{\partial f}{\partial \omega^A} = \hat{g}_A(x, \pi) - g_A(x, \pi) \tag{16.2}$$

for some $\hat{g}_A, g_A$ so that, from the linearised transition relation given in the question,

$$\hat{\omega}^A + \lambda \hat{g}^A(x, \pi) = \omega^A + \lambda g^A(x, \pi). \tag{16.3}$$

This gives us the $O(\lambda)$-correction to (16.1):

$$\omega^A = -i x^{AA'} \pi_{A'} - \lambda g^A(x, \pi) \tag{16.4}$$

and a similar expression for hatted quantities. To find the metric, first take an infinitesimally neighbouring deformed line with $\delta x^{AA'} = V^{AA'}$:

$$\delta \omega^A = -i V^{AA'} \pi_{A'} - \lambda V^{CC'} \nabla_{CC'} g^A(x, \pi). \tag{16.5}$$

The conformal metric arises from the condition imposed on $V^{AA'}$ if $\delta \omega^A$ is to vanish at say $\pi_{A'} = \alpha_{A'}$. To investigate this condition, we use the style of argument which led to (10.23): from (16.2)

$$
\begin{aligned}
\pi^{A'} \nabla^{AA'} \hat{g}_B &= \pi^{A'} \nabla_{AA'} g_B \\
&= h_{AA'BB'}(x) \pi^{A'} \pi^{B'}
\end{aligned}
\tag{16.6}
$$

since the first equality defines a global, homogeneity-2 function on L_x, which must therefore be a quadratic polynomial in π. If $V^{AA'}$ in (16.5) is to be a null vector whose primed spinor part is $\alpha^{A'}$ then

$$V^{AA'} = \sigma^{AA'}{}_{BB'} \beta^B \alpha^{B'} \tag{16.7}$$

in terms of the 'connection symbols' $\sigma^{AA'}{}_{BB'}$. (These appear because all indices in this question are concrete and not abstract.) We have

$$\sigma^{AA'}{}_{BB'} = \delta_B{}^A \delta_{B'}{}^{A'} + O(\lambda)$$

so substituting (16.7) into (16.5) and requiring this expression to vanish when $\alpha_{A'} \pi^{A'} = 0$ we find

$$\sigma^{AA'}{}_{BB'} = \delta_B{}^A \delta_{B'}{}^{A'} - i\lambda h_{BB'}{}^{AA'}.$$

Finally the linearised metric satisfies

$$g_{ab} \sigma^a{}_{CC'} \sigma^b{}_{DD'} = \epsilon_{CD} \epsilon_{C'D'}$$

from which it follows that

$$g_{ab} = \eta_{ab} + 2i\lambda h_{ab} \qquad (16.8)$$

where h_{ab} is determined by the original $f(\omega, \pi)$ by (16.2) and (16.6). It remains to be checked that this h_{ab} is a potential for the linearised Weyl spinor of the linearised metric (16.8).

c) One approach is to introduce a simple coordinate and tetrad system as follows: choose any two fibres in $\mathcal{T}$ and coordinatise a holomorphic section by the values X^A, Y^A of ω^A in these two fibres. Observe that these coordinates are all null and that the metric can be written $2H_{AB}dX^A dY^B$. Now consider the holonomic tetrad

$$e^0 = \frac{\partial}{\partial X^0}, \; e^1 = \frac{\partial}{\partial X^1}, \; e^2 = \frac{\partial}{\partial Y^0}, \; e^3 = \frac{\partial}{\partial Y^1}.$$

From the geometric definition of the connection show that

$$\nabla_{e^0} e^0 = \nabla_{e^0} e^1 = \nabla_{e^1} e^0 = \nabla_{e^1} e^1 = 0$$

and similarly for (2,3), and that

$$\nabla_{e^0} e^2 = \nabla_{e^2} e^0$$

and so on. Now it is possible to deduce that the torsion vanishes.

Appendix

The GHP Equations

The spin coefficients and GHP operators have been defined in (13.17), (13.19) and (13.20). The components of curvature in the chosen spinor dyad are labelled:

$$
\begin{array}{ll}
& \text{weight} \\
\psi_0 = \psi_{ABCD} o^A o^B o^C o^D & (4,0) \\
\psi_1 = \psi_{ABCD} o^A o^B o^C \iota^D & (2,0) \\
\psi_2 = \psi_{ABCD} o^A o^B \iota^C \iota^D & (0,0) \\
\psi_3 = \psi_{ABCD} o^A \iota^B \iota^C \iota^D & (-2,0) \\
\psi_4 = \psi_{ABCD} \iota^A \iota^B \iota^C \iota^D & (-4,0) \\
\\
\phi_{00} = \phi_{ABA'B'} o^A o^B \bar{o}^{A'} \bar{o}^{B'} & (2,2) \\
\phi_{01} = \phi_{ABA'B'} o^A o^B \bar{o}^{A'} \bar{\iota}^{B'} & (2,0) \\
\phi_{02} = \phi_{ABA'B'} o^A o^B \bar{\iota}^{A'} \bar{\iota}^{B'} & (2,-2) \\
\phi_{10} = \phi_{ABA'B'} o^A \iota^B \bar{o}^{A'} \bar{o}^{B'} & (0,2) \\
\phi_{11} = \phi_{ABA'B'} o^A \iota^B \bar{o}^{A'} \bar{\iota}^{B'} & (0,0) \\
\phi_{12} = \phi_{ABA'B'} o^A \iota^B \bar{\iota}^{A'} \bar{\iota}^{B'} & (0,-2) \\
\phi_{20} = \phi_{ABA'B'} \iota^A \iota^B \bar{o}^{A'} \bar{o}^{B'} & (-2,2) \\
\phi_{21} = \phi_{ABA'B'} \iota^A \iota^B \bar{o}^{A'} \bar{\iota}^{B'} & (-2,0) \\
\phi_{22} = \phi_{ABA'B'} \iota^A \iota^B \bar{\iota}^{A'} \bar{\iota}^{B'} & (-2,-2) \\
\Lambda = \frac{1}{24} R & (0,0).
\end{array}
$$

The GHP equations form three groups. First the equations relating the curvature to derivatives of the spin coefficients:

$$
\eth\rho - \eth'\sigma = (\rho - \bar{\rho})\tau + (\bar{\rho}' - \rho')\kappa - \psi_1 + \phi_{01} \tag{A.1}
$$

$$
\þ\rho - \eth'\kappa = \rho^2 + \sigma\bar{\sigma} - \bar{\kappa}\tau - \tau'\kappa + \phi_{00} \tag{A.2}
$$

163

$$\text{þ}\sigma - \eth\kappa = \sigma(\rho + \overline{\rho}) - \kappa(\tau + \overline{\tau}') + \psi_0 \tag{A.3}$$

$$\text{þ}\tau - \text{þ}'\kappa = \rho(\tau - \overline{\tau}') + \sigma(\overline{\tau} - \tau') + \psi_1 + \phi_{01} \tag{A.4}$$

$$\eth\tau - \text{þ}'\sigma = -\rho'\sigma - \overline{\sigma}'\rho + \tau^2 + \kappa\overline{\kappa}' + \phi_{02} \tag{A.5}$$

$$\text{þ}'\rho - \eth'\tau = \rho\overline{\rho}' + \sigma\sigma' - \tau\tau' - \kappa\kappa' - \psi_2 - 2\Lambda. \tag{A.6}$$

Next the commutators of the GHP operators acting on a quantity η of weight (p, q):

$$\begin{aligned}
(\text{þ}\text{þ}' - \text{þ}'\text{þ})\eta = {}& (\overline{\tau} - \tau')\eth\eta + (\tau - \overline{\tau}')\eth'\eta \\
& - p(\kappa\kappa' - \tau\tau' + \psi_2 + \phi_{11} - \Lambda)\eta \\
& - q(\overline{\kappa}\overline{\kappa}' - \overline{\tau}\overline{\tau}' + \overline{\psi}_2 + \phi_{11} - \Lambda)\eta \tag{A.7}
\end{aligned}$$

$$\begin{aligned}
(\text{þ}\eth - \eth\text{þ})\eta = {}& \overline{\rho}\eth\eta + \sigma\eth'\eta - \overline{\tau}'\text{þ}\eta - \kappa\text{þ}'\eta \\
& - p(\rho'\kappa - \tau'\sigma + \psi_1)\eta \\
& - q(\overline{\sigma}'\overline{\kappa} - \overline{\rho}\overline{\tau}' + \phi_{01})\eta \tag{A.8}
\end{aligned}$$

$$\begin{aligned}
(\eth\eth' - \eth'\eth)\eta = {}& (\overline{\rho}' - \rho')\text{þ}\eta + (\rho - \overline{\rho})\text{þ}'\eta \\
& + p(\rho\rho' - \sigma\sigma' + \psi_2 - \phi_{11} - \Lambda)\eta \\
& - q(\overline{\rho}\overline{\rho}' - \overline{\sigma}\overline{\sigma}' + \overline{\psi}_2 - \phi_{11} - \Lambda)\eta. \tag{A.9}
\end{aligned}$$

Finally the Bianchi identities become:

$$\begin{aligned}
\text{þ}\psi_1 - \eth'\psi_0 - \text{þ}\phi_{01} + \eth\phi_{00} = {}& -\tau'\psi_0 + 4\rho\psi_1 - 3\kappa\psi_2 + \overline{\tau}'\phi_{00} \\
& - 2\overline{\rho}\phi_{01} - 2\sigma\phi_{10} + 2\kappa\phi_{11} \\
& + \overline{\kappa}\phi_{02} \tag{A.10}
\end{aligned}$$

$$\begin{aligned}
\text{þ}\psi_2 - \eth'\psi_1 - \eth'\phi_{01} + \text{þ}'\phi_{00} + 2\text{þ}\Lambda = {}& \sigma'\psi_0 - 2\tau'\psi_1 + 3\rho\psi_2 - 2\kappa\psi_3 \\
& + \overline{\rho}'\phi_{00} - 2\overline{\tau}\phi_{01} - 2\tau\phi_{10} \\
& + 2\rho\phi_{11} + \overline{\sigma}\phi_{02} \tag{A.11}
\end{aligned}$$

$$\begin{aligned}
\text{þ}\psi_3 - \eth'\psi_2 - \text{þ}\phi_{21} + \eth\phi_{20} - 2\eth'\Lambda = {}& 2\sigma'\psi_1 - 3\tau'\psi_2 + 2\rho\psi_3 - \kappa\psi_4 \\
& - 2\rho'\phi_{10} + 2\tau'\phi_{11} + \overline{\tau}'\phi_{20} \\
& - 2\overline{\rho}\phi_{21} + \overline{\kappa}\phi_{22} \tag{A.12}
\end{aligned}$$

$$\begin{aligned}
\text{þ}\psi_4 - \eth'\psi_3 - \eth'\phi_{21} + \text{þ}'\phi_{20} = {}& 3\sigma'\psi_2 - 4\tau'\psi_3 + \rho\psi_4 - 2\kappa'\phi_{10} \\
& + 2\sigma'\phi_{11} + \overline{\rho}'\phi_{20} - 2\overline{\tau}\phi_{21} \\
& + \overline{\sigma}\phi_{22} \tag{A.13}
\end{aligned}$$

and the contracted Bianchi identities are

$$
\begin{aligned}
\text{\th}\phi_{11} + \text{\th}'\phi_{00} - \text{\dh}\phi_{10} - \text{\dh}'\phi_{01} + 3\text{\th}\Lambda &= (\rho' + \bar{\rho}')\phi_{00} + 2(\rho + \bar{\rho})\phi_{11} \\
&\quad - (\tau' + 2\bar{\tau})\phi_{01} - (2\tau + \bar{\tau}')\phi_{10} \\
&\quad - \bar{\kappa}\phi_{12} - \kappa\phi_{21} + \sigma\phi_{20} \\
&\quad + \bar{\sigma}\phi_{02} \qquad\qquad\qquad \text{(A.14)} \\
\text{\th}\phi_{12} + \text{\th}'\phi_{01} - \text{\dh}\phi_{11} - \text{\dh}'\phi_{02} + 3\text{\dh}\Lambda &= (\rho' + 2\bar{\rho}')\phi_{01} + (2\rho + \bar{\rho})\phi_{12} \\
&\quad - (\tau' + \bar{\tau})\phi_{02} - 2(\tau + \bar{\tau}')\phi_{11} \\
&\quad - \bar{\kappa}'\phi_{00} - \kappa\phi_{22} + \sigma\phi_{21} \\
&\quad + \bar{\sigma}'\phi_{10}. \qquad\qquad\quad \text{(A.15)}
\end{aligned}
$$

A great economy of notation is achieved since many of these equations are actually four equations; the original, the complex conjugate and the primes of these. Thus for example the commutator $(\text{\th}\text{\dh} - \text{\dh}\,\text{\th})$ tacitly includes the complex conjugate $(\text{\th}\text{\dh}' - \text{\dh}'\,\text{\th})$ and the primes $(\text{\th}'\text{\dh}' - \text{\dh}'\,\text{\th}')$ and $(\text{\th}'\text{\dh} - \text{\dh}\,\text{\th}')$.

Bibliography

Atiyah, M.F., Drinfeld, V., Hitchin, N. and Manin, Y. (1978) Phys. Lett. **A65** 185

Atiyah, M.F. and Ward, R.S. (1977) Comm. Math. Phys. **55** 117

Bailey, T.N. (1985) Proc. Roy. Soc. Lond. **A397** 143

Bailey, T.N. and Baston, R.J. eds (1990) *Twistors in Mathematics and Physics*, LMS Lecture Notes **156**, CUP, Cambridge

Bailey, T.N., Ehrenpreis, L. and Wells, R.O. (1982) Proc. Roy. Soc. Lond. **A384** 403

Bartnik, R. (1989) Phys. Rev. Lett. **62** 2346

Baston, R.J. (1984) Twistor Newsletter **17**

Baston, R.J. and Eastwood, M.G. (1989) *The Penrose Transform: Its Interaction With Representation Theory*, OUP, Oxford

Baston, R.J. and Mason, L.J. (1987) Class. Quant. Grav. **4** 815–826

Besse, A. (1987) *Einstein Manifolds,* Springer-Verlag, Berlin

Bott, R. and Tu, L. (1982) *Differential Forms in Algebraic Topology*, Springer-Verlag

Burns, D. (1979) in *Complex Manifold Techniques in Theoretical Physics*, eds. D. Lerner and P.D. Sommers, Pitman, London

Chern, S.S. (1967) *Complex Manifolds Without Potential Theory*, van Nostrand Reinhold, New York

Corrigan, E., Fairlie, D., Yates, R. and Goddard, P. (1978) Comm. Math. Phys. **58** 223

Curtis, W.D., Lerner, D.E. and Miller, F.R. (1978) J. Math. Phys. **19** 2024–2027

Do Carmo, M.P. (1976) *Differential Geometry of Curves and Surfaces*, Prentice-Hall

Do Carmo, M.P. (1992) *Riemannian Geometry*, Birkhauser

Donaldson, S.K. and Kronheimer, P.B. (1990) *Geometry of Four-Manifolds*, Clarendon Press, Oxford

Dougan, A.J. (1991) *Definitions of Quasi-Local Momentum*, D.Phil. thesis: Oxford University

Dougan, A.J. and Mason, L.J. (1991) Phys. Rev. Lett. **67** 2119–2122

Eastwood, M.G. (1981) Proc. Roy. Soc. Lond. **A374** 431–445

Eastwood, M.G. (1983) Twistor Newsletter **15**

Eastwood, M.G. and Ginsberg, M.L. (1981) Duke Math. J. **48** 177

Eastwood, M.G., Penrose, R. and Wells, R.O. (1981) Comm. Math. Phys. **78** 305

Eastwood, M.G. and Pilato, A. (1991) Pac. J. Math. **151** 201–215

Eastwood, M.G. and Tod, K.P. (1982) Math. Proc. Camb. Phil. Soc. **92** 317–330

Field, M . (1982) *Several Complex Variables and Complex Manifolds*, CUP

Geroch, R. (1968) J. Math. Phys. **9** 1739

Geroch, R., Held, A. and Penrose, R. (1973) J. Math. Phys. **14** 874

Griffiths, P.A. and Harris, J. (1978) *Principles of Algebraic Geometry*, Wiley, New York

Gross, L. (1964) J. Math. Phys. **5** 687-695

Grothendieck, A. (1957) Am. J. Math. **79** 121

Gunning, R.C. (1966) *Lectures on Riemann Surfaces*, Princeton U.P.

Gunning, R.C. and Rossi, H. (1965) *Analytic Functions of Several Complex Variables*, Prentice-Hall

Hansen, R.O., Newman, E.T., Penrose, R. and Tod, K.P. (1978) Proc. Roy. Soc. Lond. **A363** 445

Hawking, S.W. and Ellis, G.F.R. (1973) *The Large-Scale Structure of Space-Time*, CUP

Henkin, G.M. and Manin, Yu. I. (1980) Phys. Lett. **B95** 405

Hicks, N.J. (1965) *Notes on Differential Geometry*, van Nostrand Reinhold

Hirzebruch, F. (1966) *Topological Methods in Algebraic Geometry*, Springer-Verlag

Hitchin, N.J. (1979) Math. Proc. Camb. Phil. Soc. **85** 465

Hitchin, N.J. (1982) Comm. Math. Phys. **83** 579

Hitchin, N.J. (1983) Comm. Math. Phys. **89** 145

Hodges, A.P. (1982) Physica **A114** 157

Hodges, A.P. (1983a) Proc. Roy. Soc. Lond. **A385** 207

Hodges, A.P. (1983b) Proc. Roy. Soc. Lond. **A386** 185

Hodges, A.P. (1985a) Proc. Roy. Soc. Lond. **A397** 341

Hodges, A.P. (1985b) Proc. Roy. Soc. Lond. **A397** 375

Hodges, A.P., Penrose, R. and Singer, M.A. (1989) Phys. Lett. **B216** 48

Hormander, L. (1966) *An Introduction to Complex Analysis in Several Variables*, van Nostrand Reinhold

Horowitz, G.T. and Perry, M.J. (1982) Phys. Rev. Lett. **48** 371–374

Huggett, S.A. (1990) in *Twistors in Mathematics and Physics*, eds. T.N. Bailey and R.J.Baston, CUP

Huggett, S.A. (1992) Class. Quant. Grav. **9** 127–135

Huggett, S.A. ed (1994) *Twistor Theory*, Marcel Dekker

Huggett, S.A. and Singer, M.A. (1990) Twistor Newsletter **30**

Huggett, S.A. and Singer, M.A. (1991) Trans. Amer. Math. Soc. **324** 41

Hughston, L.P. (1979) *Twistors and Particles*, Springer Lecture Notes in Physics 79

Hughston, L.P. and Hurd, T.R. (1981) Proc. Roy. Soc. Lond. **A378** 141

Hughston, L.P. and Hurd, T.R. (1983a) Phys. Rep. **100** 273

Hughston, L.P. and Hurd, T.R. (1983b) Phys. Lett. **B127** 201

Hughston, L.P. and Ward, R.S., eds. (1979) *Advances in Twistor Theory*, Pitman

Isenberg, J., Yasskin, P.B. and Green, P.S. (1978) Phys. Lett. **B78** 462

Jeffryes, B.P. (1984) in *Asymptotic Behaviour of Mass and Space-Time*, ed. F.J. Flaherty, Springer Lecture Notes in Physics **202**

Jeffryes, B.P. (1986) Class. Quant. Grav. **3** L9–L12

Jones, P. and Tod, K.P. (1985) Class. Quant. Grav. **2** 565–577

Kelly, R.M., Tod, K.P. and Woodhouse, N.M.J. (1986) Class. Quant. Grav. **3** 1151–1167

Ko, M., Ludvigsen, M., Newman, E.T. and Tod, K.P. (1981) Phys. Rep. **71** 51

LeBrun, C.R. (1982) Proc. Roy. Soc. Lond. **A380** 171

LeBrun, C.R. (1984) Trans. Amer. Math. Soc. **284** 601

LeBrun, C.R. (1991) J. Diff. Geom. **34** 223–253

Ludvigsen, M. and Vickers, J.A.G. (1982) J. Phys. **A15** L67–L70

Ludvigsen, M. and Vickers, J.A.G. (1983) J. Phys. **A16** 1155–1168

• Manin, Yu. I. (1988) *Gauge Field Theory and Complex Geometry*, Springer-Verlag

Mason, L.J. and Hughston, L.P. (1990) *Further Advances in Twistor Theory*, Pitman

Merkulov, S.A. (1991) Class. Quant. Grav. **8** 2149–2162

Merkulov, S.A. (1992a) Func. An. Appl. **26**, 89–90

Merkulov, S.A. (1992b) Class. Quant. Grav. **9** 2369–2394

Misner, C.W., Thorne, K.S. and Wheeler, J.A. (1973) *Gravitation*, Freeman, San Francisco

Morrow, J. and Kodaira, K. (1971) *Complex Manifolds*, Holt, Rinehart and Winston

Nester, J.M. (1981) Phys. Lett. **A83** 241–242

Newman, E.T. (1976) Gen. Rel. Grav. **7** 107

Palais, R.S. ed. (1965) *Seminar on the Atiyah-Singer Index Theorem*, Princeton U.P.

Penrose, R. (1955) Math. Proc. Camb. Phil. Soc. **51** 406

Penrose, R. (1956) Math. Proc. Camb. Phil. Soc. **52** 17

Penrose, R. (1965a) Proc. Roy. Soc. Lond. **A284** 159

Penrose, R. (1965b) in *Relativity, Groups and Topology*, eds. B. and C.M. deWitt, Gordon and Breach

Penrose, R. (1966) in *Perspectives in Geometry and Relativity*, ed. B. Hoffmann, Indiana U.P.

Penrose, R. (1967) J. Math. Phys. **8** 345

Penrose, R. (1968a) Int. J. Theor. Phys. **1** 61

Penrose, R. (1968b) in *Battelle Recontres*, eds. C.M. deWitt and J.A. Wheeler, Benjamin, New York

Penrose, R. (1969) J. Math. Phys. **10** 38

Penrose, R. (1972) *Techniques of Differential Topology in Relativity*, S.I.A.M., Philadelphia

Penrose, R. (1974) in *Group Theory in Non-Linear Problems*, ed. A.O. Barut, Reidel, Dordrecht

Penrose, R. (1975) in *Quantum Gravity : An Oxford Symposium*, eds. C.J. Isham, R. Penrose and D.W. Sciama, OUP

Penrose, R. (1976) Gen. Rel. Grav. **7** 31

Penrose, R. (1982) Proc. Roy. Soc. Lond. **A381** 53

Penrose, R. (1983) in *The Mathematical Heritage of H. Poincaré*, Proc. Symp. Pure Math. 39, Am. Math. Soc.

Penrose, R. (1984) in Twistor Newsletter **18**

Penrose, R. (1986) in *Gravitation and Geometry*, eds. W. Rindler and A. Trautman, Bibliopolis, Naples

Penrose, R. (1992) in *Gravitation and Modern Cosmology*, eds. A. Zichichi, N. de Sabbata and N. Sanchez, Plenum Press, New York

Penrose, R. and MacCallum, M.A.H. (1972) Phys. Rep. **C6** 242

Penrose, R. and Rindler, W. (1984 & 1986) *Spinors and space-time*, vols. 1 & 2, CUP

Penrose, R., Sparling, G.A.J. and Tsou, S.T. (1978) J. Phys. **A11** L231

Penrose, R. and Ward, R.S. (1980) in *Einstein Centennial Volume of the Gen. Rel. Grav. Society*, ed. A. Held, Plenum Press

Pirani, F.A.E. (1965) in *1964 Brandeis Summer Institute: Lectures on General Relativity*, Prentice-Hall

Reula, O. and Tod, K.P. (1984) J. Math. Phys. **25** 1004–1008

Sachs, R.K. (1961) Proc. Roy. Soc. Lond. **A264** 309

Shaw, W.T. (1983) Proc. Roy. Soc. Lond. **A390** 191

Singer, M.A. (1990) Comm. Math. Phys. **133** 75

Spanier, E.H. (1989) *Algebraic Topology*, Springer-Verlag

Sparling, G.A.J. (1975) in *Quantum Gravity : An Oxford Symposium*, eds. C.J. Isham, R. Penrose and D.W. Sciama, OUP

Spivak, M. (1975) *Differential Geometry*, Publish or Perish

Tod, K.P. (1977) Rep. Maths. Phys. **11** 339

Tod, K.P. (1980) Surveys in High Energy Physics **1** 267

Tod, K.P. (1982) Math. Proc. Camb. Phil. Soc. **92** 331

Tod, K.P. (1983a) Proc. Roy. Soc. Lond. **A388** 457

Tod, K.P. (1983b) Proc. Roy. Soc. Lond. **A389** 369

Tod, K.P. (1984) Gen. Rel. Grav. **16** 435

Tod, K.P. (1986) Class. Quant. Grav. **3** 1169–1188

Tod, K.P. (1990a) in *Twistors in Mathematics and Physics*, eds. T.N. Bailey and R.J. Baston, CUP

Tod, K.P. (1990b) in *Geometry of Low-Dimensional Manifolds*, vol. 1, eds. S.K. Donaldson and C.B. Thomas, CUP

Tod, K.P. (1992) in *Recent Advances in General Relativity*, eds. A.I. Janis and J.R. Porter, Birkhauser, Boston

Tod, K.P. and Perjes, A. (1976) Gen. Rel. Grav. **7** 903

Tod, K.P. and Ward, R.S. (1979) Proc. Roy. Soc. Lond. **A368** 411

Ward, R.S. (1977) Phys. Lett. **A61** 81

Ward, R.S. (1978) Proc. Roy. Soc. Lond. **A363** 289

Ward, R.S. (1980) Comm. Math. Phys. **78** 1

Ward, R.S. (1981) Comm. Math. Phys. **79** 317

• Ward, R.S. and Wells, R.O. (1990) *Twistor Geometry and Field Theory*, CUP

Wells, R.O. (1980) *Differential Analysis on Complex Manifolds*, Springer-Verlag, New York

Wells, R.O. (1981) Comm. Math. Phys. **78** 567

Witten, E. (1978) Phys. Lett. **B77** 394

Witten, E. (1981) Comm. Math. Phys. **80** 381–402

Woodhouse, N.M.J. (1983) Phys. Lett. **94A** 269

Woodhouse, N.M.J. (1985) Class. Quant. Grav. **2** 257

Woodhouse, N.M.J. (1987) Class. Quant. Grav. **4** L121–L123

Woodhouse, N.M.J. and Mason, L.J. (1988) Nonlinearity **1** 73–114

Index